Palgrave Studies in Science an

Series Editor
Sherryl Vint
Department of English
University of California
Riverside, CA, USA

This book series seeks to publish ground-breaking research exploring the productive intersection of science and the cultural imagination. Science is at the centre of daily experience in twenty-first century life and this has defined moments of intense technological change, such as the Space Race of the 1950s and our very own era of synthetic biology. Conceived in dialogue with the field of Science and Technology Studies (STS), this series will carve out a larger place for the contribution of humanities to these fields. The practice of science is shaped by the cultural context in which it occurs and cultural differences are now key to understanding the ways that scientific practice is enmeshed in global issues of equity and social justice. We seek proposals dealing with any aspect of science in popular culture in any genre. We understand popular culture as both a textual and material practice, and thus welcome manuscripts dealing with representations of science in popular culture and those addressing the role of the cultural imagination in material encounters with science. How science is imagined and what meanings are attached to these imaginaries will be the major focus of this series. We encourage proposals from a wide range of historical and cultural perspectives.

Advisory Board
Mark Bould, University of the West of England, UK
Lisa Cartwright, University of California, US
Oron Catts, University of Western Australia, Australia
Melinda Cooper, University of Sydney, Australia
Ursula Heise, University of California Los Angeles, US
David Kirby, University of Manchester, UK
Roger Luckhurst, Birkbeck College, University of London, UK
Colin Milburn, University of California, US
Susan Squier, Pennsylvania State University, US

More information about this series at
http://www.palgrave.com/gp/series/15760

Jean-Baptiste Gouyon

BBC Wildlife Documentaries in the Age of Attenborough

palgrave
macmillan

Jean-Baptiste Gouyon
Department of Science and Technology
Studies
University College London
London, UK

Palgrave Studies in Science and Popular Culture
ISBN 978-3-030-19981-4   ISBN 978-3-030-19982-1   (eBook)
https://doi.org/10.1007/978-3-030-19982-1

Cover image: Mint Images Limited/Alamy Stock Photo
Cover design by eStudio Calamar

This Palgrave Macmillan imprint is published by the registered company Springer Nature Switzerland AG
The registered company address is: Gewerbestrasse 11, 6330 Cham, Switzerland

To Joséphine

Acknowledgements

Work for this book started when I researched my doctoral dissertation, more than a decade ago. Since then, many people have supported my work, provided me with much needed encouragement and contributed, in one way or another, to keep this project going: to Amanda Rees, at the Department of Sociology of the University of York, for her guidance and support during my doctorate, and her unflinching trust in my ability to write a dissertation; to Nik Brown and Robin Wooffitt, who formed my thesis advisory panel; to Greg Radick and Andrew Webster, who examined my thesis and asked questions which enabled me to take this research further—to all of them, I would like to express my gratitude.

Lorraine Daston invited me to spend a year at the Max Planck Institute for the History of Science as a postdoctoral fellow. This was a most formative time, and I am grateful to her as well as Gregg Mitman, Erika Milam, Paula Amad, Fernando Vidal, Daniela Helbig, Nasser Zakariya, Monica Aufrecht, and Etienne Benson, among others, for their collegiality during my stay in Berlin.

Louise North welcomed me over the years at the BBC Written Archives Centre in Caversham Park and was most patient in providing me with all the help needed to repeatedly access the material without which this book would not exist. James McQueen, at the BBC film archives, was also most helpful, readily providing me with copies of many programmes I would not have been able to access otherwise. All BBC copyright material is reproduced courtesy of the British Broadcasting Corporation. All rights reserved.

A few historical figures of British wildlife television have shared their memories and views with me over the years. Sir David Attenborough welcomed me twice in his home; Sheila Fullom was a precious witness all along and sent me a wealth of material, some of it quite unique; Peter Jones shared his memories and took me around Bristol, giving me a sense of the wildlife film-making community there; Desmond Morris was kind enough to answer my numerous emails, as did Tony Soper, between two expeditions. I am grateful to all for our exchanges and their welcoming generosity.

I thank Tim Boon of the Science Museum in London for his feedback on some chapters and being an ever-attentive and supportive presence in my professional life. His infectious enthusiasm, extraordinary generosity, and untiring kindness are an inspiration. I thank Charlotte Sleigh for her continuous encouragements over the years. She believed in this book before I did. I thank my colleagues from the Department of Science and Technology Studies at University College London (UCL), especially Jon Agar and Edd Bankes, who provided me with feedback on some chapters, and Joe Cain for his friendly and energetic support since I joined UCL. Last but by no means least, my partner, Céline, has been there from the start, lending her eagle eyes, ignoring my grumpiness, to critically review the manuscript in the most helpful fashion despite her many other commitments. For this and much more I am deeply thankful.

Contents

1

Introduction

David Attenborough is no doubt too modest to mention his own name, but in Britain at any rate it is largely due to his own work for the BBC that the lay public has been guided gently from simple programmes about animals to complex and sometimes quite profound essays on primitive societies and the nature of human co-operation.[1]

'Pleasure. Only Pleasure.'[2]

At its peak, the audience for the 2017 BBC wildlife series *Blue Planet 2* reached 14.1 million viewers, making it the year's most watched television programme in Britain.[3] And, commentators claimed, 'many of the

---

[1] Norman Swallow, 1966, *Factual Television*, London: Focal Press, p. 154.

[2] J. Burgess, & D. Unwin, 1984, 'Exploring the living planet with David Attenborough', *Journal of Geography in Higher Education*, 8(2), 93–113, 102.

[3] Graham Ruddick, 'Blue Planet II is year's most watched British TV show', *The Guardian*, 6 November 2017. Available online at https://www.theguardian.com/media/2017/nov/06/blue-planet-ii-years-most-watched-tv-show-david-attenborough.

© The Author(s) 2019
J.-B. Gouyon, *BBC Wildlife Documentaries in the Age of Attenborough*,
Palgrave Studies in Science and Popular Culture,
https://doi.org/10.1007/978-3-030-19982-1_1

programme's marvels are new not just to television but to science itself'.[4] Given the large number of viewers wildlife television programmes command today, and the claims associated with them, these documentaries are a key part of the apparatus through which our increasingly urbanised societies obtain their knowledge of the natural world.

Wildlife documentaries fulfil a very primordial need: to know about the world we live in and the other life forms sharing it with us. Some of the earliest figurative pictorial representations humans produced some 15,000 years ago were paintings on the walls of such caves as Lascaux in south-western France and Altamira in Cantabria (Northern Spain). They mainly show animals moving, hunting, feeding, and reproducing (Azéma 2006). What were these paintings for? Nobody knows for sure. And although we cannot be certain about what they meant to those who produced them, one thing is beyond doubt: they were part of early humans' attempts at making sense of the world they inhabited, recording their knowledge of other life forms which, to them, signified life or death. Likewise, wildlife documentaries are meant to help us understand the world we inhabit and find our place in it. But, as such, they deserve to be scrutinised, for, often straddling boundaries—between nature and culture, or human and animal—they unsettle these boundaries only to re-draw them in ways which naturalise specific orders of things. For example, they use the spectacle of anthropomorphised animals, or of humans in the wild, to naturalise social norms (e.g. heteronormativity, the nuclear family as the basic unit of social structure, etc.; Chris 2006; Haraway 1989). This book explores the genealogy of wildlife television in Britain and provides readers with some elements to make sense of how television programmes and documentaries about wildlife have contributed to fashioning how we see ourselves and where we stand in the world over the past five decades. These are essential conversations to have in light of our current environmental predicaments.

Wildlife documentaries are about knowledge. They are premised on 'an epistemology which itself is grounded in the recording of the particular, physical real by camera and microphone' (Corner 1996: 3). But although,

---

[4] Elle Hunt, 'Blue Planet II: From octopus v shark to fish that crawl, the series's biggest discoveries', *The Guardian*, 10 December 2017. Available online at https://www.theguardian.com/tv-and-radio/2017/nov/20/blue-planet-ii-what-have-we-learned-so-far.

as any kind of documentary work, they have—or claim to have—evidential value, they also are intrinsically artificial. As David Attenborough noted in an interview in 1984, just after the release of *The Living Planet* (Chapter 9):

> In fact, there is precious little that is natural … in any film. You distort speed if you want to show things like plants growing, or look in detail at the way an animal moves. You distort light levels. You distort distribution, in the sense that you see dozens of different species in a jungle within a few minutes, so that the places seems [sic.] to be teeming with life. You distort size by using close up lenses. And you can equally well distort sound. What the film-maker is trying to do is to convey a particular experience in as vivid a way as he can.[5]

Wildlife documentaries contain and generate knowledge of nature for their audiences. At the same time, they obey one imperative: the necessity of spectacle (Gouyon 2016).[6] The artifice of film-making Attenborough describes in this quote is enrolled in creating a spectacular experience for viewers, one from which they will derive not only knowledge but also a sense of wonder, and ultimately pleasure. For, still in Attenborough's words: 'It seems to me that science, fundamentally, is concerned with defining man's relationship with the natural world—making sense of it. And when it does that, it brings great pleasure.'[7] The consumption of such visual spectacle and entertainment as wildlife films is a subjective experience steeped in emotions (Mitman 1999). For this reason, wildlife documentaries are traditionally conceived of as ill-suited to convey objective knowledge and educate their audiences (Shapin and Barnes 1976; Greenhalgh 1989). If wildlife documentary makers are to be able to achieve their aim, which is to be recognised as trustworthy sources of knowledge about nature, they therefore need to deploy strategies that enable them

---

[5]J. Burgess, and D. Unwin, 1984, 'Exploring the living planet with David Attenborough', *Journal of Geography in Higher Education*, 8(2), 93–113, 103.

[6]See Louson (2018) on the genealogy of wildlife documentaries' spectacular dimension. See also Mitman (1993).

[7]J. Burgess, & D. Unwin, 1984, 'Exploring the living planet with David Attenborough', *Journal of Geography in Higher Education*, 8(2), 93–113, 102.

to resolve this tension between knowledge and entertainment—evidence and artifice—which lies at the heart of the wildlife documentary.

Identity Fashioning

A way to look at the history of wildlife television is to follow the strategies wildlife broadcasters have deployed through time fashioning their public identity as reliable sources of knowledge about nature and constructing the film-making apparatus as a legitimate tool to produce knowledge of the world. A key device in this joint process of identity fashioning is the wildlife 'making-of documentary' (MOD). MODs are now a staple of every high-profile nature series, each episode ending with a ten-minute segment revealing secrets about the shooting of iconic scenes in that episode. As a subgenre of the wildlife television programme, however, MODs are quite a recent addition to wildlife broadcasters' repertoire and are not self-evident. When asked why they were made in the first place, David Attenborough provides a two-fold explanation blending epistemological concerns with preoccupations associated with the material aspects of television broadcasting. Wildlife broadcasters began producing MODs, he explains, when they needed audiences to trust them. He also hints at the fact that these ten-minute add-ons at the end of episodes are convenient to introduce some flexibility in the length of programmes. They make it easier for television executives to sell the programmes to countries where the default length of programmes is fifty minutes and not sixty.[8] However, some wildlife film-makers do think that those concluding segments are not unproblematic. Peter Jones, for instance, who was the executive producer for the third mammoth Attenborough series, *The Trials of Life* (1990), whose last episode, *Once More into the Termite Mound*, is a MOD (Chapter 9), remarked:

> now the MOD is a ten-minute segment added to the main programme without even a pause, something that I personally dislike. My understanding is that the controllers and planners find an hour-long programme much

---

[8]David Attenborough, interview with author, 25 November 2015.

more suited to their scheduling. At the time of *Trials*, the customary length for each programme in the 13-part series was precisely fifty minutes so we are really dealing with changing conventions—but I still do not like the customary current outcomes, as they undermine the momentary emotional and intellectual catharsis when you reach the end of well-constructed films.[9]

This short incursion into the topic of wildlife MODs brings up three essential facets of wildlife documentaries: (1) they are about knowledge; (2) they are commodities; (3) they are well-crafted storytelling devices.

There is nothing necessary about MODs in the first place, nor about their format: a ten-minute concluding segment or self-contained episode. A question which motivated me to research this book early on was how, when and why did the MOD become a genre in wildlife television? The answer lies in the history of wildlife television in Britain, which is what this volume explores. Although wildlife MODs ended up appearing only episodically in the book, when they do so, they stand as milestones in the slow process of accretion of the public identity of wildlife film-makers. The question of how wildlife film-makers construct their public identity, present themselves and their practice to audiences—what could be called the making-of impulse—is central to the argument. Having had a chance, several years ago, to be a science journalist, I have had a long standing desire to understand how scientific knowledge exists in the public realm and how these public forms of existence in turn contribute to shaping the public culture of science. A claim central to this volume is that looking at the history of wildlife television shows how hard British wildlife broadcasters have worked to make their practice, programme making, a part of the production of knowledge of the natural world. In doing so, they not only presented themselves as knowledge producers but also contributed to shaping our common understanding of what it means to study wild animals in the field, and what scientists' findings in the field mean to all of us.

---

[9] Peter Jones, interview with author, 22 August 2014.

Summary of the Argument

Wildlife MODs and other forms of disclosures are part of the broader history of wildlife film-makers' public fashioning of their social identity. Wildlife film-makers are best seen as scientific showmen. And as such, they are highly concerned with issues of disclosure and concealment. As historian of science, Iwan Morus, wrote about Victorian scientific showmen:

> Deciding what to show and what to hide was an issue that concerned anyone involved in the business of exhibition. The key to successful performance often lay in the management of information between performer and audience. … Even if showmen did not want their audiences to see all their secrets, they wanted them to see enough to recognise and applaud the skill and ingenuity that lay behind the successful show. Strategies of concealment and exposure therefore lay at the heart of scientific performers' self-fashioning. (Morus 2006: 105)

This volume approaches the history of wildlife documentaries from this perspective of broadcasters' self-fashioning of their social identity as truth tellers. In doing so, it identifies three stages in the history of wildlife television, punctuated by as many MODs. First, as wildlife television making was becoming a profession, patrolled by experts in the recreation of nature on-screen—broadcasters—the latter found it necessary to disclose some of their techniques to justify their use of the artifice of film-making whilst remaining true to nature. A film like *Unarmed Hunters* (BBC 1963), arguably the first wildlife MOD made for British television, defines wildlife broadcasters as professionals capable of producing natural historical knowledge through film-making. This trajectory entailed distancing wildlife television from the culture of amateur natural history in which early nature broadcasting originates (Chapter 2) and moving closer to scientists investigating animal behaviour in the field (Chapter 6). A driving force for this rapprochement between scientists and broadcasters was David Attenborough, who started his career at the BBC, producing natural history programmes in collaboration with Julian Huxley and then the London Zoo (Chapters 3 and 4). But, as scientists got more involved in wildlife television, broadcasters felt the necessity to distinguish their profession from that of scientists and to assert that if scientists wanted to become film-makers, they had to stop being scientists (Chapter 5). Yet, scientists-turned-film-makers' involvement with wildlife film-making had a transformative effect on the practice of filming wildlife for television (Chapter 7). The film-makers of Oxford Scientific Films turned film craft, the capacity to design ingenious film-making devices to show natural phenomena from hitherto unseen perspectives, into the equivalent of amateur naturalist cameramen's field craft. Camerawork became a key component of the stories told in wildlife television programmes, as exemplified in the 1972 *The Making of a Natural History Film* (Chapter 8).

Through this trajectory, the meaning of objectivity in relation to wildlife television is transformed, as is the film-making apparatus's relationship with nature. In the first period of wildlife television, dominated by amateur naturalists, support for the films' claims to objectivity came from film-makers providing evidence that they had not interfered with nature when filming, simply holding the camera to passively record their encounter with the wild and share it with audiences (Gouyon 2011a). But once the production of wildlife television has become the work of professionals, claims to objectivity for the films they present rest on the idea of 'mechanical objectivity' (Daston and Galison 2007). The images appearing on-screen stand as the outcome of a process which does not involve a direct encounter between a human subject and nature but one between mechanical devices and nature. Film-makers' intervention consists in creating the conditions for this interaction, and ultimately results in the creation of nature. For, the version of nature offered to viewers, presented as revealing hitherto unknown aspects of the natural world, only exists there, on-screen. With the professionalisation of wildlife film-making, wildlife documentaries become performative in the sense that the film-making process creates the subject of the documentary. Nature becomes the outcome of filming.[10]

Life on Earth (1979) came out of this professionalisation and exemplifies the performative capacity of wildlife documentary making (Chapter 9). With this series, wildlife film-makers do not anymore simply show nature; they make it visible. Broadcasters' close relationship with scientists was

---

[10]For more on the performative documentary, see Bruzzi (2006).

essential to produce this pinnacle of the history of wildlife television, as they relied on scientists to find out where and when to go and film the stories they wanted to tell to depict the evolution of life on the planet. *Life on Earth*, the first of the major so-called Attenborough series—an enduring genre to this day—ushered the figure of the 'telenaturalist' in the professional culture of wildlife television. I first proposed this term in my PhD dissertation to emphasise the distinctive character of Attenborough's screen performances (Gouyon 2011a, b). By contrast with studio hosts presenting natural history magazines, such as Peter Scott, whose natural history expertise exists outside the studio, the telenaturalist's expertise proceeds from television and depends on the medium to manifest itself. The telenaturalist is a television-show character who performs natural history on the TV screen for no other purpose than the production of television programmes, and who relies on the film-making apparatus to produce natural history knowledge. This performance is a one-man show that rests, partly, on the purposeful exclusion of other human beings from the picture, notably scientists. Yet, given the cognitive authority our culture vests in science, the telenaturalist's trustworthiness, and his appearing as an epistemic equal to scientists, ultimately depends on him being seen as standing on an equal footing with scientists. This is achieved with wildlife MODs produced alongside *Life on Earth*, *The Living Planet*, and *The Trials of Life* (Chapter 9).

Beyond presenting film-makers as professionals and disclosing their technical virtuosity, wildlife MODs define participation in film-making as a shared means for telenaturalists and scientists to collaboratively investigate the workings of nature. Submitting to the constraints of film-making in the field, they all obtain new knowledge and understanding of the natural world. MODs justify delegating the act of knowing to the film-making apparatus—cameras, editing room, dubbing studio—and turn wildlife television into a mechanically objective tool of knowledge production. *Once More into the Termite Mound*, the MOD for *The Trials of Life* (BBC 1990), places field scientists in relation to film-making in a position comparable to that of the telenaturalist. Emphasising the film-making apparatus's central epistemic function over human subjects' agency, MODs are also displays of modesty which reinforce the telenaturalist's status as a truth teller. A question left hanging at the end of the book is whether the epithet 'telenaturalist' can apply to other wildlife television screen personalities, or will remain indissolubly tied to the age of Attenborough.

The Age of Attenborough

When preparing this book, I met twice with David Attenborough. My hope was to find out from the man himself answers to all the questions raised by repeated dives into the material kept at the BBC Written Archives Centre (BBCWAC) in Caversham Park, Reading. On my visits to Richmond, I met the most affable of men, at the same time fully aware of his status and unassuming about it. Two memories will stay with me for a long time: the moment I rung the bell, Attenborough opened his door in one decisive pull, as if lifting a curtain, darting his inquisitive blue eyes at me whilst introducing himself. Upon my second visit, his daughter welcomed me and lead me to the sitting room, and I stood on the doorstep, waiting for Attenborough to reach the end of a piece from the Romantic repertoire he was magisterially interpreting on a grand piano. Perhaps more significant for a historian, though, is that in Attenborough I also found someone who considered that he had written everything he wanted to say about his own involvement in the history of wildlife television in Britain in his memoirs (Attenborough 2010), and was somewhat reluctant to talk about himself any further. Above all, Attenborough came across as very forward-looking, not so much interested in the past as he is in the present and what his next projects will be. Although he left no question unanswered, nothing motivated him more than discussing his then current endeavour: the filming of a series on bioluminescence. To meet with David Attenborough when one studies the history of wildlife television in Britain is more than meeting the individual; it is getting in touch, by proxy, with all the people who appear in this book, from Julian Huxley and Desmond Hawkins to Mick Rhodes and Christopher Parsons, and accessing the world view which informed the development of wildlife broadcasting in Britain during the second half of the twentieth century.

French sociologist Pierre Bourdieu described the ways in which individuals involved in a field of activity like scientific research achieve status and recognition through the accumulation of symbolic and material resources.

For a scientist, symbolic resources could be academic titles (e.g. honorary doctorates and medals) and material ones could be grants, a laboratory, people working in their research team. The sum of these resources Bourdieu called symbolic capital. Within a given field, the more symbolic capital individuals accumulate, the more power and authority they gain, which in turn enables them to shape their field to their advantage and accumulate even more material and symbolic resources, and eventually achieve dominance in the field (Bourdieu 1994, 2001).

BBC Wildlife Documentaries in the Age of Attenborough is a historical study of the development of a field, that of wildlife television in Britain, roughly over the five decades that followed the Second World War. As the chapters in this book show, the production of wildlife television programmes and documentaries is a profoundly collective process, only successfully achieved through well-coordinated team work. However, wildlife television is also a field in which recognition goes to those demonstrating the property of skill to produce good television programmes, as judged by audience reception and ratings. In this field, symbolic capital is unequally divided. Those individuals who produce programmes successful with large audiences will accumulate more of it than those only able to address a niche segment of the television watching population. And as in any field, those who accumulate more symbolic capital are able to shape the field so that it will enable them to accumulate even more capital. Within a given field of human activity, some individuals have therefore more influence than others.

In the field of wildlife television, David Attenborough managed early on to produce television programmes which chimed well with the cultural context and consistently achieved very good ratings. In other words, Attenborough was very successful early on in his career at accumulating symbolic capital, and over the years has accumulated more than anyone else in his field. Accordingly, he can be said to have shaped the field to his advantage, making it work so that his way of approaching wildlife television making became the norm. The title for this volume therefore is intended to reflect Sir David's position of dominance in the field and to acknowledge his epoch-making influence on it. But it should not be taken to imply that this book is a biographical account of David Attenborough. It is not. Instead, *BBC Wildlife Documentaries in the Age of Attenborough* tries to understand how wildlife television as it was produced during a given epoch—which goes from the early 1950s to the early 2000s and perhaps will be remembered as the age of Attenborough—has come to stand as the accepted stereotype of how nature should appear on-screen today.

Animals, Television and Natural History

The history of television partly is a history of animals on-screen.[11] Animals, as conveniently self-moving objects, figured prominently in early programmes broadcast in Britain between 1936 and 1939, participating in the effort to fashion what was then a new medium as a truthful source of factual knowledge. A medium from the outset framed as dealing with factual information as opposed to mere entertainment, television needed to demonstrate its trustworthiness. As with early cinema so with television, animals became 'central figures in the presentation of [a] new and "progressive" technology' (Burt 2001: 206); animal programmes, dispensing expert knowledge about wildlife to viewers, served to establish television's epistemic credentials.

Natural history is a pursuit born out of a desire to document, to catalogue the world, to identify the unknown, and in this way to know ourselves (Ritvo 1997; Burt 2001; Razac 2002; Daston and Mitman 2005). Since the nineteenth century, natural history has enjoyed the status of a universal discipline, broad enough in its topics of interests from minerals to insects and plants, to appeal to the many. A reservoir of knowledge and practices from which professional scientific disciplines keep emerging (Outram 1996), natural history is also an enduring scientific training ground. At the interface between professional and amateur science, it was and still is considered well adapted to convey the basic principles of scientific enquiry, for curiosity can spring out of the observation of familiar objects. For example, Grant Allen, a late-nineteenth-century populariser

---

[11] For a specific study of animals on television, see Mills (2017).

of evolution, reckoned that although not all non-scientists would be interested in the minutiae of the evolution of anatomical structures or physiological processes, starting from such common objects as berries, shells or tadpoles, and explaining their general features was an efficient way of presenting general evolutionary principles to a wide and diverse audience (Lightman 2009: 219). Natural objects are familiar ones from which unfamiliar stories, based on scientific claims, can be derived. These stories are not simply educative; they are also edifying. For to teach people how to order the world is also to teach them how to position themselves in relation to others in the social world (Ritvo 1987).

Placing us in a close visual proximity with animals, to peer into the intimacy of their existence, wildlife documentaries allow us to experience genuine alterity. They alleviate our feeling of existential solitude. Animals, the other living beings we share our finite environment with, are our ontological partners in 'the arts of living and dying well in multispecies symbiosis' (Haraway 2015: vii–viii). Besides, wildlife documentaries are a source of emotional comfort, for they relieve our feeling of 'ontological insecurity', the 'sense of confusion, loss, unpredictability and anxiety' which sociologist Adrian Franklin (1999: 56) associates with late modernity's characteristic 'perpetual state of change' and lack of direction. Wildlife films offer us, notably with the so-called blue-chip documentaries (Bousé 2000), the spectacle of timeless, unchanging nature. From the 1950s to the early 2000s, the period covered in this book, wildlife consistently appeared on television as both object of knowledge and subject for the material performance of television production. But whilst this duality persisted over the period, the presentation of wildlife on television has itself been inflected by two factors: the establishment of television as an institution capable of generating an authoritative discourse on the world and, within the institution, the rise and development of a professional culture distinct from both radio and cinema, that places a premium on broadcasters' ability to show things and to tell stories, principally by visual means.

UCL, April 2019.

References

Attenborough, D. (2010). *Life on air: Memoirs of a broadcaster*. London: Random House.

Azéma, M. (2006). La représentation du mouvement au Paléolithique supérieur: Apport du comparatisme éthographique à l'interprétation de l'art pariétal. *Bulletin de la Société préhistorique française, 103*(3), 479–505.

Bourdieu, P. (1994). *Raisons pratiques. Sur la théorie de l'action*. Paris: Editions du Seuil.

Bourdieu, P. (2001). *Science de la science et réflexivité*. Paris: Raisons d'Agir.

Bousé, D. (2000). *Wildlife films*. Philadelphia: University of Pennsylvania Press.

Bruzzi, S. (2006). *New documentary*. London: Routledge.

Burt, J. (2001). The illumination of the animal kingdom: The role of light and electricity in animal representation. *Society & Animals, 9*(3), 203–228.

Chris, C. (2006). *Watching wildlife*. Minneapolis: University of Minnesota Press.

Corner, J. (1996). *The art of record: A critical introduction to documentary*. Manchester: Manchester University Press.

Daston, L., & Galison, P. (2007). *Objectivity*. New York: Zone Books.

Daston, L., & Mitman, G. (Eds.). (2005). *Thinking with animals: New perspectives on anthropomorphism*. New York: Columbia University Press.

Franklin, A. (1999). *Animals and modern cultures*. London, Thousand Oaks, and New Delhi: Sage.

Gouyon, J.-B. (2011a). From Kearton to Attenborough: Fashioning the telenaturalist's identity. *History of Science, 49*(1), 25–60.

Gouyon, J.-B. (2011b). The BBC natural history unit: Instituting natural history film-making in Britain. *History of Science, 49*(4), 425–451.

Gouyon, J.-B. (2016). 'You can't make a film about mice just by going out into a Meadow and looking at mice': Staging as knowledge production in natural history film-making. In M. Willis (Ed.), *Staging science* (pp. 83–103). London: Palgrave Pivot.

Greenhalgh, P. (1989). Education, entertainment and politics: Lessons from the great international exhibitions. In P. Vergo (Ed.), *The New Museology* (pp. 74–98). London: Reaktion Books.

Haraway, D. (1989). *Primate visions*. London: Routledge.

Haraway, D. (2015). Cosmopolitical critters: Preface for cosmopolitan animals. In K. Nagai, et al. (Eds.), *Cosmopolitan animals* (pp. vii–xiv). London: Palgrave Macmillan.

Lightman, B. (2009). *Victorian popularizers of science: Designing nature for new audiences.* University of Chicago Press.

Louson, E. (2018). Taking spectacle seriously: Wildlife film and the legacy of natural history display. *Science in Context, 31*(1), 15–38.

Mills, B. (2017). *Animals on television: The cultural making of the non-human.* London: Palgrave Macmillan.

Mitman, G. (1993). Cinematic nature: Hollywood technology, popular culture, and the American Museum of natural history. *Isis, 84*(4), 637–661.

Mitman, G. (1999). *Reel nature: America's romance with wildlife on film.* Cambridge, MA: Harvard University Press.

Morus, I. R. (2006). Seeing and believing science. *Isis, 97*(1), 101–110.

Outram, D. (1996). New spaces in natural history. In N. Jardine, et al. (Eds.), *Cultures of natural history* (pp. 249–265). Cambridge: Cambridge University Press.

Razac, O. (2002). *L'écran et le zoo. Spectacle et domestication, des expositions coloniales à Loft Story.* Paris: Denoël.

Ritvo, H. (1987). *The animal estate: The English and other creatures in the Victorian age.* Cambridge, MA: Harvard University Press.

Ritvo, H. (1997). *The platypus and the mermaid and other figments of the classifying imagination.* Cambridge MA: Harvard University Press.

Shapin, S., & Barnes, B. (1976). Head and hand: Rhetorical resources in British pedagogical writing, 1770–1850. *Oxford Review of Education, 2*(3), 231–254.

2

Live from Alexandra Palace, Wildlife Comes to Television

'Always Start with the Familiar': Early Wildlife TV Before World War Two

Wildlife film-making is the art of making nature visible. On 25 January 1937, a slightly atypical wildlife television programme introduced early television viewers to some practical aspects of wildlife film-making. *The Making of Documentary and 'Secrets of Nature' Films*, lasting fifteen minutes, featured Mary Field (1896-1968), then the celebrated producer of the British instructional *Secrets of Nature* (1922–1933) and *Secrets of Life* (1934–1950) film series. She explained the 'special methods by which these films [were] made, the apparatus used', partially lifting the veil on cinema's power of illusion.[1] To illustrate her talk, Field showed sequences from the two *Secrets* series. The programme itself was live, and so did not survive in archival form. But from an accompanying piece published in the *Radio Times* in the last week of January 1937, it is possible to get a sense of what was shown and discussed. Sequences taken from films depicting roots or runner beans growing, plankton swimming around in a rock pool

---

[1] *Radio Times*, 'Television supplement', 22 January 1937, p. 6. On the *Secrets* series see Boon (2008).

© The Author(s) 2019
J.-B. Gouyon, *BBC Wildlife Documentaries in the Age of Attenborough*,
Palgrave Studies in Science and Popular Culture,
https://doi.org/10.1007/978-3-030-19982-1_2

and insect larvae developing introduced viewers to such techniques as time-lapse, micro- and macro-photography. *The Making of Documentary and 'Secrets of Nature' Films* drew viewers' attention to the necessary very technical nature of wildlife film-making.

The clips shown in the TV programme were meant to highlight the amount of time necessary and the practical hurdles film-makers had to overcome—such as the lethal overheating of fragile life forms under the intense lighting required to film in close-up—if the films were to reveal the workings of nature. Disclosing some of the techniques film-makers deployed made it clear to viewers that they were watching the result of much work and ingenuity. Through the construction of dedicated sets, and the use of elaborate devices and pieces of equipment, film-makers produced truthful reconstructions of nature. From these, Field claimed, spectators could obtain knowledge of usually hidden natural phenomena. Her presentation of the films from the *Secrets of Nature* and *Secrets of Life* series as technical achievements asserted the necessity of artifice when bringing nature on-screen, a theme still running through wildlife film-makers' discourse to this day.

As the producer of a series of nature films, Field insisted that audiences were at the core of her endeavour. Telling stories that made the unfamiliar appear mundane when presenting natural history subjects, was indispensable, she said, if viewers were to relate to the films at all.

> Makers of nature films need to remember that many of the public are not much interested in semi-instructional films unless the subjects are familiar to them. … A golden rule for making nature films is this: always start with the familiar, and never let members of your audience feel that they have strayed from the paths of their ordinary experience.[2]

Here Field acknowledged the importance of the commentary, and its contribution to steering viewers' experience toward the familiar, notably through repeatedly anchoring strange images in everyday lived experience. The commentary on pictures of 'microscopic salt-water life' would thus always remind audiences 'of the familiar appearance of the sea-shore at low

---

[2] *Radio Times*, 'Television supplement', 22 January 1937, p. 5.

tide', presenting microscopic scenes as seaside pictures, 'not remote laboratory studies'.[3] Commentary in a nature film is not just about imparting knowledge, or information. It participates in the mediation of nature, in the construction of meaning through visual means, insofar as it directs viewers' perception. The commentary's power of allusion is key for audiences to suspend their disbelief and for the film's illusionary power to work.

Concluding her piece in the *Radio Times*, Mary Field remarked: 'The two all-important requisites in this kind of film—variety and reality—sound so simple to attain. But their attainment demands all the skill and all the resources of the film-maker'.[4] This early programme encapsulates a key feature of wildlife television: wildlife broadcasters' desire to both reveal how nature works and, simultaneously, how *they* work, drawing the attention to how these representations have been obtained. This dual endeavour turns the production of wildlife television into a key element of the programmes, as much as what the programmes are ostentatiously about: nature. To paraphrase media theorist Marshall McLuhan (1964), wildlife television is as much about wildlife as it is about itself; the medium is, at least partially, the message. This book explores how those involved in the production of wildlife television developed their skill and advertised it, thereby supporting the claim that they could be trusted as reliable sources of knowledge of the natural world.

In the 1930s, visual culture was dominated by cinema, and early television was perceived at first as necessarily standing in competition with it. To create a space for television in the media landscape where it could co-exist peacefully with cinema, proponents of the new medium presented it as fact based, as opposed to cinema, defined as fiction based, the realm of illusion. Thus Cecil Lewis, one of the co-founders of the BBC noted in 1937:

I do not believe [television] will conflict with the cinema or the theatre; as broadcasting did, it will develop its own technique. … I believe that its unique feature, in which it differs from any other form of entertainment

---

[3] *Radio Times*, 'Television supplement', 22 January 1937, p. 5.
[4] *Radio Times*, 'Television supplement', 22 January 1937, p. 5.

or news service, is in its ability to bring the actuality before the public at the very moment it is happening.[5]

Animals were enrolled in this demarcation effort. Whereas cinema-goers were more likely to encounter them on-screen as part of fictional depictions of adventurous explorations in tropical lands, when first featured on television, animals came along as deeply embedded within a discourse of zoological expertise. As the first week of television transmission in Britain was coming to a close, *The Zoo Today*, the very first programme featuring live animals on television was broadcast, on Saturday 7 November 1936. In the studio was David Seth-Smith (1875–1963), the curator of mammals and birds at the London Zoo. For its subsequent instalments, Seth-Smith's programme changed its title to *Friends from the Zoo*, and until the summer 1939, when television transmission came to a halt because of World War Two, the series occurred at least once a month, which made it the most regular animal programme in early British television. A live broadcast, it offered viewers the entertaining prospect of animals getting out of hand. In 1936, Seth-Smith was already known to audiences from a series of popular talks on animals he had delivered on the radio since 1932. His reputation in natural history circles had grown from spending six months in Australia in 1907, collecting some 700 animals there on behalf of the Zoological Society of London (ZSL), which awarded him its silver medal. Seth-Smith's television programmes were little more than a televised version of his radio shows. He introduced animals from the London Zoo's collection from a studio at Alexandra Palace, aided by keepers from the Zoo. On a couple of occasions, BBC cameras went to the Zoo itself, vicariously guiding viewers between cages and enclosures, with Seth-Smith delivering the running commentary. Be it in the studio or at the London Zoo, Seth-Smith was an expert, dispensing certified zoological knowledge to his viewers through the mediation of television which thus stood as a trustworthy source of facts about the natural world.

When television broadcasting resumed, on 7 June 1946, following the interruption caused by World War Two, wildlife also returned on the screen. At first, things did not appear to have changed: *Friends from the*

---

[5]Cecil Lewis, 'Television, out-of-doors', *Radio Times*, 8 January 1937, p. 5.

Zoo was broadcast on Saturdays as had been the case until 1939. The concept was the same, only the presenter was different. Geoffrey Vevers had replaced David Seth-Smith as the superintendent of the zoological gardens, and so had become the new host for the programme, similarly introducing zoo animals from a studio at Alexandra Palace, with the help of zoo keepers.

But in the early 1950s two new approaches to presenting wildlife on TV emerged, each informed by a different cultural repertoire. One originated in Bristol. Involving celebrity naturalist Peter Scott, it was rooted in the culture of amateur natural history. The other, elaborated in London, featured celebrity wildlife film-makers from the 1930s, Armand and Michaela Denis, and extolled the culture of imperial big game hunting. These two cultures were two aspects of the Victorian culture of natural history, itself enmeshed in British imperialism and a nascent culture of mass consumption. Natural history, an observation-based investigation of the natural world, developed as an overwhelmingly visual culture in the early modern period (Long 2002; Bleichmar 2012). It became tightly associated with the British imperial project. The production and circulation of visual representations of exotic nature promoted and legitimated the colonial exploitation of the natural world (Browne 1996). In parallel, the establishment of places such as zoos and museums of natural history encouraged the consumption of visual representations of the natural world as a form of rational entertainment for the middle classes emerging in the nineteenth century (Fyfe and Lightman 2007; Bensaude-Vincent and Drouin 1996). The co-existence, in the early BBC wildlife output, of Scott's amateur naturalist programmes and the Denises' exotic big game films, reflects competing understandings within the Corporation of the form wildlife television should take, and what its aims should be, educational (Scott) or entertaining (Denis). In both cases, though, wildlife television was less visibly tied to a discourse of zoological expertise and its trustworthiness had to be established by other means.

Look with Peter Scott: Bringing Amateur Natural History to Television

Although the war had brought television broadcasting to a stand-still, it had created meeting opportunities for the people who would eventually build the foundations of wildlife television. Desmond Hawkins (1908–1999), for example, worked for the BBC as a freelance producer on the *Daily War Report* programme. After the war ended, he edited the *BBC War Report* anthology (1946). Through this work he met Frank Gillard (1908–1998), who had been a war correspondent and had become head of BBC West programmes in 1945.[6] In 1946, Gillard appointed Desmond Hawkins as the only staff radio feature producer in Bristol, with the specific task of producing natural history programmes for amateur naturalists and listeners with a general interest in wildlife. The first programme was *The Naturalist*, which principally featured bird songs captured in the wild by Ludwig Koch (1881–1974), a German refugee and pioneer of wildlife recording (Guida 2018). *The Naturalist*, just like *Birds in Britain* which came afterwards, and *Bird Song of the Month*, typically mixed studio sequences with 'sound-pictures' taken in field locations. Popular with audiences, these programmes attracted collaborations from such recognised naturalists of the period as Peter Scott (1909–1989), Maxwell Knight (1900–1968) and Maurice Burton (1898–1992), who bestowed on them their expertise. Here were individuals who could be trusted to deliver technologically mediated yet truthful representations of nature, from which audiences could derive reliable natural historical knowledge. But most importantly, these programmes positioned the BBC in Bristol within the regional network of amateur natural history, creating the substrate from which wildlife television would grow from 1953 onward (Davies 2000).

The opening act of British wildlife television after World War Two was a live outside broadcast on 1 May 1953, from Slimbridge, the cottage which naturalist Peter Scott had bought in Gloucestershire to establish his

---

[6]Leonard Miall, 'Obituary: Desmond Hawkins', *The Independent*, 8 May 1999. Available online at https://www.independent.co.uk/arts-entertainment/obituary-desmond-hawkins-1092138.html. Last accessed 5 March 2019.

Wildfowl Trust. The idea for the outside broadcast with Peter Scott, took shape in the autumn of 1952 after Desmond Hawkins and his assistant, Tony Soper (1929–), had seen Peter Scott, then a regular contributor to *The Naturalist*, give a talk at Colston Hall in Bristol. To illustrate his talk, Scott drew wildfowl on stage, using large sheets of paper on an easel. A trained artist, he had spent three years, from 1927 to 1930, at Trinity College, Cambridge, reading Natural Sciences, Zoology, Botany, Physiology, and Geology (Scott 1966: 51). But having obtained his degree, he chose to carve himself a career as a wildlife painter rather than becoming a life scientist in a laboratory. Scott's a posteriori justification was that his interest lay in studying live animal behaviour, a topic which he could not then have investigated as a scientist. In the early 1930s, he noted in his memoirs: 'The science of animal behaviour had scarcely begun. To know about live animals was something less than science' (Scott 1966: 78). And so, Scott turned to drawing and painting as means of exploring the living world, in reaction against what he felt was too restrictive a view on wild animals and what counted as knowledge of them. In particular, Scott found painting superior to zoological enquiry when it came to investigate how birds move, for painting enabled the artist to produce composite images. Several representatives of a species could be put side by side in different postures on the same canvas. What appeared as a flock of birds was an artifice that enabled Scott to visually analyse and understand the movement of an individual bird through the air (Gouyon 2011).

Scott conceived of his natural history paintings and drawings as objective representations that were true to nature. But their truthfulness largely depended on his own subjective knowledge of the birds: 'Other artists did not know them [wildfowl] quite as I knew them' (Scott 1966: 97). Scott's knowledge, in the first place sensual, originated from countless hours of observing birds 'at dawn or dusk or moonlight, or in storm or frost or snow (Scott 1966: 97). With his paintings and drawings, Scott hoped to transmit a sense of his experience of being in nature and to move spectators 'in the same way as [he] was when [he] watched the flight of the wild geese, and heard their music' (Scott 1966: 97). Natural history painting is helpful to think with when considering the practice of filming nature in the wild and the materiality of films. Instead of conceiving of films as a series of discrete snapshots whose rapid succession before the eye produces the illusion of life, natural history films can be seen, just like Scott's paintings, as composite images. They are assemblages of several views of the same natural objects, which offer more to see than a single observation would (Gouyon 2011).

On their way back from Colston Hall, Hawkins and Soper discussed the televisual potential of Scott's talk. Shortly thereafter, Hawkins approached Scott to ask whether he would consider performing a similar talk on television. On 1 May 1953, Desmond Hawkins, Frank Gillard and Tony Soper arrived at Scott's cottage in Slimbridge with a BBC mobile control unit to broadcast live the first natural television programme made in Bristol. In this programme, entitled *Severn Wildfowl*, Scott gave a tour of his water birds collection and was then interviewed by Frank Gillard in his study, swiftly drawing as he talked. To supplement the whole, Scott showed a few film sequences of wildfowl from his own catalogue, commenting on the circumstances of their making. *Severn Wildfowl* combined two performances in one programme: that of the naturalist, seen painting and drawing whilst presenting his films, and that of the broadcaster, whose mastery of the outside broadcasting equipment enabled viewers to vicariously stand in the room as witnesses of the naturalist's performance. The combination of these two performances contributed to establishing television making as a way of producing natural historical knowledge. Peter Scott's first appearance on television could have taken the form of him simply showing one of his films. That this first natural history television programme was a live outside broadcast indicates that early on, broadcasters in Bristol believed that wildlife television should happen on the medium's own terms and not be a mere instance of television conveying somebody else's images. Scott's subsequent appearances similarly involved production techniques specially developed for the television medium.

At the same time as he was launching natural history on air, Hawkins was also experimenting with the broadcasting of theatre plays. Some of the broadcasting techniques these allowed him to develop were poured back into Scott's next programmes. Following the success of the first live transmission from his Slimbridge cottage, Scott started appearing monthly in live programmes filmed on a set reproducing Scott's study at home, complete with a fireplace and an easel. Designed by Desmond Chinn, the usual set designer for the BBC televised theatre plays, it was first installed in a studio in Lime Grove, London, before moving to Bristol in 1955. The choice to use such a domestic setting suggests that viewers for these programmes were mostly recruited in social groups whose pursuit of natural historical knowledge was a form of rational entertainment, and for whom the space dedicated to such enquiry was an extension of the domestic space (Shapin 1988). These were the bourgeois heirs of the early natural philosophers, gentlemen of leisure who did not take up science as a profession but placed a premium on amateurism, for the absence of material interest in the pursuit of knowledge guaranteed their status as truth tellers (Shapin 1994). Each programme would open with Scott swiftly sketching a bird in front of the camera to introduce the topic of the day. He would then present sequences of some film he had shot whilst on a natural history expedition, narrating it with a voice-over, ad-lib commentary describing the images, telling spectators what they should see.

Scott's travels were voyages of discovery in the tradition of late eighteenth- and early nineteenth-century scientific expeditions to map the biogeography of the British Empire (Browne 1996), and the films documenting them were scientific records. A case in point is *The Pink-Feet of Iceland*, broadcast on 24 June 1954, in which Scott presented the films shot during his 1951 Icelandic expedition to study the migration of pink-footed geese. The birds breed at the foot of the Hofsjökull icecap in central Iceland each spring and early summer. They migrate in the autumn to overwinter in Britain. The expedition had been intended to start ringing birds of this large Icelandic colony in the hope that some would be caught again in the winter in Britain, thus allowing Scott to evaluate the proportion of birds migrating from Iceland to Britain. A second expedition took place in July 1953 to complete the work begun in 1951. As Scott explained to readers of *The Times* in an article sent from Iceland,

The main objects of our study are to find out how many of these geese there are in the world, how many are breeding in this great colony, how many die each year, and how many young are produced. These figures can be ascertained by a system of marking and subsequently re-sampling when the marked birds have had time to mix freely with the rest of the population. …

only in 1951, with our first visit … did Iceland emerge as almost certainly the main breeding headquarters, and this particular colony of over 2000 nests as the largest known to science.[7]

The Times presented Scott's Icelandic expedition to its readers as a scientific one, which rested on the application, in the field, of the sampling technique Scott had developed at the Severn Wildfowl Trust. Scott's ambition when buying Slimbridge had been to create a research station to produce knowledge about birds and their behaviour in the wild. There, he had hired a few biologists to work for him and, to give a measure of his hopes, had offered Konrad Lorenz, the Austrian founder of ethology, to come and act as head of research for 'nine hundred pounds per year, plus use of Scott's boat the *Beatrice*' (Burkhardt 2005: 357). Although, as his obituarist remarked, 'Peter [Scott] never claimed to be an academic of any kind' (Shackleton 1989), his employing scientists at Slimbridge and during his expeditions to conduct research work on bird behaviour positioned him as a patron of the sciences, locating the Severn Wildfowl Trust within the cultural space of science. Originating in such a venture, the film sequences which Scott showed in *The Pink-Feet of Iceland* enabled viewers to witness the context of the work as well as the implementation of Scott's sampling technique. Through his commentary, Scott added enough circumstantial details to convince the audiences that the work had been conducted in the way claimed, therefore reinforcing the evidential value of the films (Shapin 1994). Finally, having with him in the studio other participants in the expedition was to produce witnesses who could vouch for the truth of his account. Thus, in *The Pink-Feet of Iceland* Scott had his wife, whom he had married during their 1951 trip to Iceland, confirming his narration of the events as well as sharing her personal experience with viewers and so enabling them to form a more detailed mental image of the expedition.

Soon, though, Scott ran out of films and turned to his naturalist friends, inviting them to join him on the set of his television show to share their own footage and narrate their expeditions. Following Scott's programme on the Icelandic geese, ornithologists James Fisher and Roger Tory Peterson came to discuss their exploration across the North American continent,

---

[7] Scott, 'An Icelandic journey. Study of Pinkfoot geese', *The Times*, 21 July 1953, p. 5.

'studying and filming wild life from the swamps of Florida to the islands of the Bering Sea'.[8] Other programmes featured footage of badgers taken at night by Ernest Neal, or close-ups of birds by Field Marshall Lord Alanbrooke, who was also the chairman of the Severn Wildfowl Trust. These programmes contributed to integrating what was to become the BBC's Natural History Unit (NHU) in existing networks of natural history (Davies 2000). They also contributed to creating the idea that a special kind of naturalist had emerged in Britain during the 1950s, 'the BBC Naturalist' (Hawkins 1957), to whom filming wildlife and appearing on television to talk about it was part of the practice of natural history.

Each of the programmes Scott hosted followed the same loose chore-ography. He would open with an introduction of the topic and then join his guest seated on an armchair. The conversation would then turn to explaining the technicalities involved in obtaining the footage, often showing photographs of the arrangement. For instance, in the programme *Badgers*, broadcast on 20 November 1954, Ernest Neal and his acolyte, practising biologist Humphrey Hewer, explained how they had progressively habituated badgers to intense lighting during night-time so that they could film them. They showed photographs of the lamps suspended in the trees above badgers' sets to illustrate their narration, bringing to the studio actual examples of the various light bulbs they had used, from the standard 25 watts to the football-sized 1000 watts lamp. They also demonstrated the 16 mm camera they had used to obtain the images shown in the programme, explaining how they had muffled the sound of the camera's mechanism so as not to disturb the badgers. Such displays of equipment, and disclosures of methods of filming alongside the resulting footage, reinforced the evidential value of the films, the film-makers' cognitive authority, and the overall trustworthiness of the television programme. Humorous acknowledgements of failures and mistakes during the film-making process, either in the shape of mishandlings of equipment, animal subjects not behaving as expected, or comical disruptions by strangers, further supported the film-maker's claims to trustworthiness (Shapin 1994).

---

[8] *Radio Times*, 2 July 1954, p. 44.

The breakthrough for Scott's programmes came on 15 January 1955 when he invited German film director Heinz Sielmann to show his film *Woodpeckers*. A feat of ingenuity, the film, which Scott had first seen at the 1954 International ornithology symposium in Basel, revealed what was happening inside a woodpecker nest. Sielmann, who was working for the Munich Institute for Film and Photography, had spent a year in the woods filming the birds. The genius of the movie had been to replace part of a tree trunk in which woodpeckers had nested with a glass pane, covering it with a black cloth (Sielmann 1959). This artifice enabled him to capture never-before-seen images of the intimate life of the birds as the parents entered the nest, fed the young, and took turns. The film, which, for broadcasters, was technically revolutionary, had an immense impact, the equivalent, for wildlife television, of the 1953 coronation. For the first time, the assembled audience was able to enter the intimate world of animal life and see for themselves how they behaved.

Most of the appeal of *Woodpeckers* rested in the impression it gave viewers that they stood in a privileged position from which they could gaze at the intimate life of birds. The film cast the motion picture camera as a passive, unproblematic recording device mediating the film-maker's observations, a tool to penetrate secret places of nature and reveal hitherto hidden truths. During his conversation with Scott in front of the TV cameras, Sielmann mobilised familiar tropes of naturalists' discourse of trustworthiness, narrating the troubles he had gone through to obtain the images, how it had tested his patience, endurance, physical resistance, and his ingenuity. The unassuming modesty of his discourse and general demeanour worked as performative understatements that presented Sielmann as a 'modest witness', who stood to gain nothing from being believed or not. This performance of disinterestedness worked to reinforce his trustworthiness, signalling to viewers that they were free to withhold their assent to the claims laid in the film (Shapin and Schaffer 1985). On the day following transmission, everybody was talking about *Woodpeckers* (Attenborough 2010) and the programme propelled Bristol's natural history television front stage. Eager to build on this success, Desmond

Hawkins gave Scott a regular series 'with a title of its own that would soon become familiar and popular'.[9] That title was *Look*. The first edition aired on 14 June 1955 with, once again, Heinz Sielmann, this time presenting a film about foxes. A monthly broadcast produced by Tony Soper—the first of a long list of producers—*Look* built a solid following, consistently attracting several millions of viewers. The combination of a conversational style of presentation with the positioning of viewers as witnesses, both channelled through Peter Scott's gentlemanly figure and steeped in the traditional culture of amateur natural history, defined the series as an authorised source of expert knowledge about the natural world. In 1957, Desmond Hawkins could boast:

> Programmes like *Birds in Britain* and *Look* have shown that they can hold the attention of an audience of several millions. Such broadcasters as Peter Scott or James Fisher enjoy a measure of popularity that would certainly not be scorned by the more orchidaceous and spectacular stars of the entertainment world. (Hawkins 1957: 7)

The Social Usefulness of Wildlife Television

Hawkins, a convinced pacifist who had only joined the military during the Second World War out of concerns about Nazi Germany's military advances, was a card-carrying member of the Peace Pledge Union.[10] To him, wildlife programmes' instant success with audiences originated in a collective longing for peace and respite after the war. Interviewed in the 1982 documentary *Wildlife Jubilee* marking the twenty-fifth anniversary of the BBC's NHU, Hawkins recalled:

> In those post-war years, there was a wish, somehow to find, to regain Paradise Lost. This I think was the appeal of the wild to those first audiences—to

---

[9] Desmond Hawkins, 'The animal people', *Radio Times*, 21 March 1968, p. 32.
[10] Leonard Miall, 'Obituary: Desmond Hawkins', *The Independent*, 8 May 1999. Available online at https://www.independent.co.uk/arts-entertainment/obituary-desmond-hawkins-1092138.html. Last accessed 5 March 2019.

just switch on and see all those wonderful animals and feel free. It was a liberation.

A shared belief that the animal world held the secret of peace and could offer a remedy to human folly drove the action of students of animal behaviour in the immediate post-war period. In September 1946, the Roscoe B. Jackson Memorial Laboratory in Bar Harbor, Maine (USA) gathered a crowd of animal behaviour and animal sociology scholars for a symposium on genetics and social behaviour. Some of the proceedings appeared in 1950, as a special issue of the *Annals of the New York Academy of Sciences*. One of John Paul Scott's aim when organising this conference was 'to remedy the deplorable lack of scientific control over destructive social phenomena such as warfare, crime, and poverty' (Scott 1950: 1003). Throughout the fifties, numerous efforts to eradicate war, hunger and poverty, and to better human society, originated in the life sciences. Julian Huxley's *Humanist Frame* (1961) is another, slightly belated, manifestation of this movement.

When he began broadcasting wildlife television and radio programmes in the mid-fifties, Hawkins shared in these views. To many people, Hawkins claimed, wildlife television programmes such as *Look* bring a 'relief from everyday care and anxieties' through 'the contemplation of the natural wildlife of the world'.

> The more our man-made perplexities torment and frighten us, the more we seem to find peace and refreshment in looking steadily at the permanent conditions of life and in trying to understand a little more clearly the rules and patterns of animal existence. And here the amateur student and the scientist come to terms, with a possibility of general intelligibility and a shared objective. (Hawkins 1957: 7)

As the Cold War was settling in, with such techno-scientific innovations as atomic and hydrogen bombs and spacecraft regularly making the headlines, Hawkins promoted wildlife television as an escape from the worries associated with geopolitical turmoil. But to Hawkins, there was more to wildlife television than an antidepressant virtue. Programmes like *Look* could enrol television audiences in a scientifically informed project of social reform, whereby the scientific exploration of the animal kingdom could help establish the rules that should govern human societies for the greater good. To Hawkins and other early television broadcasters, wildlife television was a social technology, a tool to channel social-cultural norms and values and enact social change. In the fast-urbanising Britain of the 1950s, one perceived urgent issue among naturalists was the population's growing estrangement from nature. They saw the restoration of this link as essential for the betterment of British society. The development of new towns, such as Stevenage, Harlow, Welwyn Garden City and Hatfield, built to re-house populations in bombed areas, the expansion of suburban habitats, and the development of agriculture (Sandbrook 2005) had led many in the natural history community to worry about expanding urban populations' loss of connection with the natural world. It could only result in a split between urban and rural populations, leading to the destruction of local wildlife, and making any return to nature very difficult indeed. Proponents of wildlife television in the late 1950s found this perceived risk of the estrangement of urban populations from nature a justification for their enterprise, claiming that their efforts could sensitise populations about what was at stake. For example, to the eminent bird photographer G. K. Yeates, one of Peter Scott's guests in *Look*, wildlife television could mend the British nation. Television programmes celebrating the British wilderness could allow for mutual cultural understanding between those living in the cities and those inhabiting the countryside:

> The great films that have appeared on television have shown the city-dweller the existence of a world he only vaguely dreams of and could never personally experience. … if our country has any one problem that it must above all else solve, it is to bridge the gap that exists between town and country. … If the nature photographer can feel that he has in any way helped to bridge that gap, he will be well rewarded. (Yeates 1957: 20)

The perceived social usefulness of wildlife television, as well as its success with audiences, led to the creation of the NHU in July 1957. This followed internal lobbying by Frank Gillard and Desmond Hawkins with the BBC's board of management. Hawkins' argument was that formally establishing the NHU would be the logical outcome of the development and concentration of expertise in producing natural historical content, both for radio and television, which had taken place in Bristol since 1946. Further, this new entity could act as a bridge between the BBC and the scientific world. This vision manifested itself in the profile advertised for the position of head of the NHU. The ideal candidate was to be an experienced broadcaster who, at the same time, possessed high academic qualifications in the life sciences.[11] The first person to be approached for the position, as early as March 1957, was David Attenborough. A degree in Natural Sciences from the University of Cambridge, and his reputation within the BBC as a successful programme maker, notably based on his very popular *Zoo Quest* series (Chapter 3), made him the ideal candidate. Yet he declined the offer, preferring to remain posted in London.[12] Instead, Hawkins privileged an expertise in programme making, and so the position of senior producer went to a seasoned broadcaster, Nicholas Crocker, who had begun his career at the BBC as a radio outside-broadcast assistant producer in Bristol in 1946, moving on to television in 1952 as a senior outside broadcast producer. Before coming to television, Crocker had worked in commercial radio and, most significantly, as a producer and actor at the Old Vic Theatre in London, where he had acquired thorough knowledge of theatre technique, an asset for the direction of studio-based programmes. John Burrell, the director of the Old Vic Company, recommended him 'as a theatre man [with] an imaginative and vigorous personality'.[13] Crocker was highly regarded at the BBC, and was included in 1955 in the '"hard core" of skilled and experienced staff' to whom a very generous 'no escape' contract was offered after the opening of television to competition. Together with Hawkins and a few others in Bristol, he was perceived as indispensable to 'the maintenance of the present output' to the extent that the BBC 'could not face the prospect of him being enticed away by commercial

---

[11] See for example the memo 'Senior Producer, Nature Unit, West Region' from Controller West Region, 4 October 1957, BBCWAC L1/2,264/1.

[12] Deputy director of Television broadcasting (Cecil McGivern) to Director of Television broadcasting, 18 March 1957, BBCWAC Folder T31/385.

[13] John Burrell, [undated] letter of reference, BBCWAC L1/2,264/1.

television'.[14] At the head of the NHU, Crocker increased the technical standard of its output. However, not being a naturalist, he felt that his contribution could only be limited. He thus resigned after a year, arguing that further development needed the leadership of a trained naturalist. To succeed him, the BBC hired Bruce Campbell, already known to the Bristol broadcasters for his regular radio programmes on birds and for hosting the monthly children television programme *Out of Doors*. A novice in television programme production, Campbell underwent several months of training in London before stepping into his new role, where he remained until 1962. Campbell brought to the NHU his natural history connections and his academic credentials. The first full-time secretary of the British Trust for Ornithology from 1948 to 1958, a role which brought him in close collaboration with influential ornithologists such as James Fisher, Campbell also held a PhD in Comparative Ornithology from the University of Edinburgh, the first doctorate awarded in Britain, based on ornithological field work.[15]

The alternation of appointees from a seasoned broadcaster to a certified naturalist illustrates the ambition at the core of the foundation of the NHU. Desmond Hawkins was keen, from the start, to create a structure that could inhabit both the cultural space of science and that of television broadcasting, at high levels of expertise in both fields. Appraising Crocker's contribution in 1959, Hawkins noted that 'the work he has done, in instilling a proper regard for production standards, was most valuable to a newly-formed team'.[16] One of the first programmes produced in Bristol under Campbell's leadership, in which he acted as a presenter, *The Return of the Osprey*,[17] was recognised in ornithological circles as resting on research 'more comprehensive than the entire literature'. With Campbell as its editor, scientific congresses took to inviting participation from the NHU, whilst members of the NHU frequently contributed papers to scientific

---

[14]Leslie Page, Establishment Officer, Television, 'Retention of Staff in the face of Competition', memo to Controller West Region, 30 March 1955, BBCWAC L1/2,264/1.

[15]David Snow, 'Obituary: Bruce Campbell'. *The Independent*, 13 January 1993. Available online at https://www.independent.co.uk/news/people/obituary-bruce-campbell-1478202.html. Last accessed 24 October 2018.

[16]Desmond Hawkins, Annual confidential report for Nicholas Crocker, 30 September 1959, BBCWAC L1/2,264/1.

[17]Produced by Christopher Parsons and first transmitted on 23 August 1959.

journals.[18] Just as Crocker's period at the head of the NHU had led to its being recognised within the BBC as a centre of broadcasting excellence, Campbell's editorship positioned the NHU as a centre of contributory expertise to the scientific exploration of the natural world.

Building up Wildlife Television Personalities with Armand and Michaela Denis

But whilst Hawkins was busy building up natural history as the signature output of BBC West, others in London, also aware of the broadcasting value of animals and other natural history subjects, had begun appropriating the topic. Yet, Hawkins approached wildlife television as a means to create and communicate natural historical knowledge and increase viewers' awareness and appreciation of their immediate environment. By contrast, at London's Television Centre, nature and animals were, in the first instance, valued for their entertainment potential. This approach tended to privilege the spectacle of exotic charismatic megafauna over the local wildlife.

Tropical wildlife first got a boost on the television screen in October 1953, when Nairobi-based film-makers Armand and Michaela Denis made their debuts on the BBC only a few months after Hawkins had launched Scott on the air. The Denises were first featured in the BBC light entertainment programme *In Town Tonight* to promote their latest film, *Below the Sahara*, produced and distributed by the Hollywood studio RKO, and due to open in London cinemas (Denis 1966: 277). First interviewed for ten minutes, the couple explained how they worked, and then they commented on extracts from their new film *Below the Sahara*. Watching them in the studio was Cecil Madden, the flamboyant assistant to Cecil McGivern, the controller of the BBC's television service. Madden had begun his career in television in 1936, specialising in the production of live light entertainment programmes featuring ballet, drama and variety, and was well connected to the world of entertainment spectacle. David

---

[18]Desmond Hawkins, 1962, 'The BBC Natural History Unit. Report by the head of West Regional Programmes', p. 3. BBCWAC R13/462/1.

Jones, the representative of RKO in Britain, had tipped him off about the couple's visit to London, suggesting that he may be interested in meeting with them in his search for exciting new documentary material (Lewis 2007: 285). Within a few weeks of their introduction to British television viewers, the Denises had signed a contract for four further half-hour programmes, broadcast between March and May 1954. RKO had offered the BBC five-minute extracts from both *Below the Sahara* (1953), and *Savage Splendour* (1949). These were shown together with new footage the Denises had shot in Australia, in preparation for a feature film on the Great Barrier Reef, to be distributed in the UK by Ealing Studio.[19] Following these first appearances on TV, the Denises found themselves celebrities almost overnight, with people recognising them on the street. As they confided in a letter to McGivern upon their return in Nairobi: 'Even here, … we keep meeting people who recognize us, having seen us on TV in England, and their enthusiasm is a very pleasant thing for us.'[20]

Encouraged by such positive reports Madden engaged in actively constructing Armand and Michaela Denis as television personalities. In May 1954, Alan Sleath, their producer at the BBC, took the couple to a visit of the Whipsnade Wild Animal Park of the Zoological Society of London (ZSL). BBC Television's *Newsreel* filmed them; photographers from the *T.V. Mirror* were also in attendance. The Denises did not disappoint, putting on a great show. Michaela rode a three-ton rhinoceros, and Armand seemingly soothed a Bengal tiger just by talking to it.[21] Following the visit, Madden wrote to the BBC's pictorial publicity officer:

It occurs to me that the set of 26 photographs taken by us at Whipsnade with these explorers recently is so good that it might be an idea to suggest them to *the Tatler* and *the Sketch* as these might be very unusual for them. As these explorers are going to Africa shortly we could always arrange an interview so that the captions could be more vivid. The picture of Micaela [sic.] riding on a rhinoceros is very effective and definitely on the dangerous side. We are anxious to build these two up.[22]

---

[19]Madden to Programme Organiser, 8 January 1954, BBCWAC T6/112.
[20]Armand Denis to Cecil McGivern, 22 July 1954, BBCWAC T6/112.
[21]Cecil Madden, 'Filming wild animals in Africa', BBCWAC T6/112.
[22]Madden to Pictorial Publicity Officer, 31 May 1954, BBCWAC T6/112.

As they departed by plane from the UK to Nairobi, Madden arranged for news cameras to film the couple, again insisting: 'The more we can build them up as television personalities the better.'[23] 'Personality' was a key word within the BBC in the early 1950s. The number of television license holders in Britain, 3.2 million, was still quite low compared to the 10 million wireless owners. With such numbers, the television service was not financially self-sufficient. This enticed those responsible for developing this new branch of the BBC to seek ways of increasing the number of regular viewers for their programmes. Favouring the development of a personal contact between a programme and its viewers through the creation of personalities was one of these (Bennett 2011). As Paul Fox, the creator of the BBC flagship current affairs programme *Panorama* explained:

> The best way to establish the proper kind of contact is by means of a visible personality, someone who has down the years become something of a family friend, a regular visitor to the sitting room, a man whose words are respected and whose very presence has become … a guarantee of integrity and common sense. (Quoted in Swallow 1966: 63)

Wildlife was a popular topic with audiences. Building up wildlife personalities such as the Denises or Peter Scott helped draw attention to television and broaden the new medium's reach in the British population. But Madden's specialisation in light entertainment programmes meant that his approach to the task differed from Desmond Hawkins's own efforts, highlighting the coexistence, from the start, of two rival cultural repertoires within wildlife television. To Hawkins, broadcasters had to have 'a special regard for the scientific elements in Natural History programmes, even in their more popular forms', and so he endeavoured to highlight Scott's scientific credentials.[24] Peter Scott's programmes, broadcast first from his sitting room and then from a studio set representing it, firmly anchored natural history broadcasting from Bristol in the elite Victorian culture of amateur natural history. Madden, by contrast, explicitly drew from the repertoire of adventurous imperial exploration epitomised

---

[23]Madden to Alan Sleath, 4 June 1954, BBCWAC T6/112.
[24]Desmond Hawkins, 1962, 'The BBC Natural History Unit. Report by the head of West Regional Programmes', BBCWAC R13/462/1.

in such novels as Rider Haggard's *King Solomon's Mines* (1885). Steeped in the Victorian culture of imperial big game hunting, which extolled such virtues as courage and physical endurance or abnegation, explorers' accounts derived their trustworthiness from providing evidence, in the shape of trophies, that they exerted these virtues in the quest for natural knowledge (Ritvo 1987; Burrow 2013). The Denises' footage, equivalent to big game hunters' trophies, was evidence that they had behaved in the field in the virtuous way expected of imperial explorers.

The seemingly contrasting ways in which Bristol and London approached the treatment of natural history and wildlife underlines the unresolved tension between education and entertainment present in wildlife television from these early days. But these two approaches are the two sides of the same coin, for, in both cases, a cultural repertoire borrowed from the Victorian period was mobilised in support of wildlife television's claim to trustworthiness. The co-existence of the Denises with Peter Scott on British television suggests a willingness on the part of the BBC to capitalise on wildlife to attract a variety of audiences across the social spectrum. As personalities where becoming central to the budding ecology of wildlife television, popular appeal became the measure of success. When Desmond Hawkins noted in 1957 that Peter Scott enjoyed 'a measure of popularity that would certainly not be scorned by the more orchidaceous and spectacular stars of the entertainment world' (Hawkins 1957: 7), he may well have been referring to the Denises, seeing Scott's success at attracting viewers as a vindication of his approach. Earlier on, still anxious to assert Scott's popularity, Hawkins reminded readers from the *Radio Times* that he could 'engage the Festival Hall—and fill it—when nobody else had risked such a gamble'.[25] In January 1954, Scott had given two very successful public lectures in London's Royal Festival Hall, with a 2500-seat capacity, presenting films from his expeditions to South America and Iceland. This prompted Madden to invite Denis, shortly after the couple's three initial broadcasts, to emulate Scott: 'An idea for you occurs to me for the future. Peter Scott filled the Festival Hall with a lecture and some films. When you have some new films you might do the same

---

[25]Desmond Hawkins, 'Peter Scott: Lover of Wildlife', *Radio Times*, 4 March 1955, p. 9.

there.'[26] Madden was eager that the Denises—at least, Armand—could compare favourably with the amateur naturalist from Slimbridge.

Broadcast on the same platform, potentially to similar audiences, these two approaches to wildlife television mutually shaped one another and their key performers. In October 1954, the Denises signed a contract with the BBC for a series of eight films to be transmitted over eight months. The first episode of *Filming Wild Animals* aired on 6 November 1954, establishing a formula that would endure for more than a decade under varying titles. The episode opened with the two explorers introducing the storyline in a studio. The film itself was a mixture of wildlife footage, and sequences staging the couple's encounters with animals in the wild or at their home base in Nairobi with such pets as a young elephant or rhino. Armand Denis provided an ad lib voice-over commentary, with Michaela Denis only delivering short lines. The programmes consistently brought forward the two presenters' physical and affective proximity to wildlife. In his *Radio Times* billing for the series, Madden emphasised this intimate, emotional connection with animals: 'They never take or touch a gun and the most menacing wild animals seem almost literally to eat out of their hands'.[27] Pre-war wildlife films, shot for the theatre circuit, predominantly participated in the big game hunting culture, capitalising on the dangers of the chase to generate excitement in audiences. Armand Denis had made a name for himself there. Julian Huxley considered his 1938 *Dark Rapture*, documenting Denis's joint expedition with big game hunter Theodore Roosevelt across the Belgian Congo, as the 'finest' film on African wildlife (Mitman 1999: 188). After World War Two, as they attempted to launch their career on television, a medium consumed in the domestic space, the Denises emphasised instead 'the family life of animals and the wonders of nature'.[28] This approach was also in keeping with the post-war longing 'for paradise lost' which Desmond Hawkins highlighted as the cultural trend explaining the popularity of wildlife television in 1950s Britain. In Madden's words

---

[26]Cecil Madden to Armand Denis, 27 May 1954, BBCWAC T6/112.

[27]Cecil Madden, 'Filming wild animals in Africa', *Radio Times*, 29 October 1954, p. 7.

[28]Tom Walshe, 'Des Bartlett obituary', *The Guardian*, 9 November 2009. Available online at https://www.theguardian.com/environment/2009/nov/09/des-bartlett-obituary. Last accessed 25 October 2018.

The Denises' cameras open up the living world of the animal kingdom, the plant, bird and insect life of the Continent of Africa in a new and personal way. There is no doubt that they love animals and have the greatest respect for them.[29]

Madden's insistence on the couple's love and respect for animals, using arguments related to sensuality, emotion and the affective—values more akin to the culture of amateur natural history—can be interpreted as a move to emulate Peter Scott, who was similarly building on his sensual approach to wildfowl, as a painter, to support his claims to trustworthiness as a source of knowledge of the natural world. Yet, Armand and Michaela Denis lacked the kind of natural history credentials that a Peter Scott could muster. And so, rather than having the couple ape Scott, Madden forced the contrast with the gentleman of leisure, constructing the Denises as glamorous characters in adventure films, whose legitimacy as sources of knowledge of the natural world derived from their intimacy with African wildlife. Michaela Denis was instrumental in this respect. She was, as Madden reminded *Radio Times* readers, 'an explorer in her own right'. Nonetheless, 'with her red hair and her shapely trousers, [she] certainly adds a touch of glamour to the jungle scene in darkest Africa, since she invariably takes her make-up along with her'.[30]

Through his efforts to shape the Denises into television personalities, Madden imported into wildlife television some of the codes of entertainment cinema, not least the notion that spectatorship entails a willing suspension of disbelief. In so doing, Madden blurred the boundary between fact and fiction, a distinction asserted by Peter Scott's programmes. *Filming Wild Animals* belongs in the travelogue genre and narrates the two heroes' explorations and quests for remarkable animals to film, such as the female black rhinoceros Gertie, supposedly famous for the length of her horn. Along the way, encounters with fierce lions or with 'exotic people' added thrill and intrigue to the adventure. Sequences depicting the couple filming in the field, with either or both seen pointing a camera at animals, alternate with footage of wildlife purporting to be those shot by

[29]Cecil Madden, 'Filming wild animals in Africa', *Radio Times*, 29 October 1954, p. 7.
[30]Cecil Madden, 'Filming wild animals in Africa', *Radio Times*, 29 October 1954, p. 7.

the two explorers. Combined, these sequences tell a story of courageous film-making in the midst of risky adventures. However, as one commentator remarked in the magazine *Punch*, after a few episodes of *Filming Wild Animals* had aired:

> When Armand and Michaela are roughing it, taking risks and braving hardships, I doubt whether it is wise or strictly honest to omit all reference to the *other* cameramen, those who photograph hero and heroine in their hide, wading through filthy swamp or challenging rogue elephants to do their worst.

> Sooner or later every viewer realizes that he is seeing his picture not through the eye of Armand's camera but through that of some unseen and anonymous extra, someone whose burden and trials must be just as heavy as those of the stars. And when this happens the illusion is shattered and suspicion is aroused.[31]

The question of whose eye viewers were seeing nature through was crucial for establishing the trustworthiness of early wildlife television. In Peter Scott's programmes, the camera holder was clearly identified as Scott's guest, introducing a film record of the many hours spent in the field, patiently observing wild animals behaving naturally. This format negated the artifice of film-making in the sense that the camera was denied an agency in the production of the visual display. Films shown in this context were not the outcome of encounters between the camera and nature but between camera-holders and nature. The latter did not perform their trustworthy identity as wildlife film-maker through their appearance in the film but through their discussing it in Scott's studio. In the Denises' films, most footage of the couple in the wild and of wildlife was shot by Des Bartlett, a young debuting cameraman the Denises had met in Australia in 1952 and brought back with them to Nairobi. For more than ten years, Bartlett was the Denises' main camera operator before going to work for the East Anglia television series *Survival* (Chapter 4) after Armand Denis retired in 1966. Denis, a film entrepreneur whose career had begun in entertainment cinema in the 1930s, produced television programmes as

---

[31] Bernard Hollowood, 'On the air', *Punch*, 1 June 1955, p. 692.

he directed films for theatre release. This involved coordinating the work of a pair of (usually anonymous) cameramen and editing together their footage to illustrate a pre-set narrative centred on Armand and Michaela's adventures in the wild. These films were not so much about animals and nature as they were performances for the Denises to stage their identity as adventurous wildlife film-makers. Yet, this use of cinematic narrative artifices developed in the realm of fiction obfuscated the identity of the camera holder, standing in contradiction to the conventions of wildlife television as defined in Peter Scott's programmes. In the long run, this feature of the Denises' work would prove a weakness for their standing within wildlife television.

But early critics at the BBC focused, at first, less on the film-makers' trustworthiness and more on the programmes' entertainment quality. Seymour de Lobinière, then the BBC's acting controller of television programmes, judged the overall style naïve and lacking in professionalism. In his opinion the couple talked too much; Michaela Denis was 'twee and coy' and did not deliver her lines professionally enough. The programme did not create enough suspense and lacked 'a spice of danger, or semi-danger, especially for Michaela Denis'. Lastly, 'the pets atmosphere' did not go down well either, as it looked like a repeat of what viewers were already getting from programmes such as George Cansdale's *Looking at Animals*, which, in 1951, became the successor to *Friends from the Zoo* and likewise rested on displays of animals from the London Zoo.[32] However, audiences welcomed the new programmes. The Denises' producer, Alan Sleath, used their popular appeal to fend off criticisms, arguing that the programme's consistently high ratings were the sign that critics had misconceptions about audiences' tastes and expectations.[33] This ambivalence at the BBC toward the Denises would endure throughout their collaboration with the Corporation. However, as they were enjoying a promising start on television, their producer wrote to warn them of a competing attempt at bringing natural history to television audiences. Sleath was referring to

---

[32] Notes on Armand & Michaela Denis "Filming in Africa"—No. 1 seen 20th December 1954, BBCWAC T6/112.

[33] Handwritten note on 'Notes on Armand & Michaela Denis "Filming in Africa"—No. 1 seen 20th December 1954', BBCWAC T6/112.

David Attenborough's first *Zoo Quest* series, whose genesis is discussed in the next chapter.

Conclusion

From the outset, two approaches to bringing animals on-screen co-existed in wildlife television, one with its roots in the cultural repertoire of amateur natural history and focusing on British wildlife, the other an offshoot of the culture of imperial big game hunting and resting on the spectacle of exotic charismatic megafauna. The co-existence, early on, of these two approaches illustrates the notion that a tension between edification and entertainment, respectively embodied in the format of the natural history lecture and the adventure film, stands at the root of wildlife television.

Neither of these two approaches originated in television, though. Both were imported from pre-existing media: radio and cinema, respectively. The rest of this volume examines how, as television broadcasters constructed a professional culture for the new medium, those creating wildlife programmes developed a third culture of wildlife film-making, aligned neither with amateur natural history nor with big game hunting but with the scientific exploration of the world.

References

Attenborough, D. (2010). *Life on air: Memoirs of a broadcaster*. London: Random House.

Bennett, J. (2011). *Television personalities: Stardom and the small screen*. Abingdon: Routledge.

Bensaude-Vincent, B., & Drouin, J.-M. (1996). Nature for the people. In N. Jardine, et al. (Eds.), *Cultures of natural history* (pp. 408–425). Cambridge: Cambridge University Press.

Bleichmar, D. (2012). *Visible empire: Botanical expeditions and visual culture in the Hispanic Enlightenment*. Chicago: University of Chicago Press.

Boon, T. (2008). *Films of facts*. London, New York: Wallflower press.

Browne, J. (1996). 'Biogeography and empire'. In N. Jardine, et al. (Eds.), *Cultures of natural history* (pp. 305–321). Cambridge: Cambridge University Press.

Burkhardt, R. W. (2005). *Patterns of behavior: Konrad Lorenz, Niko Tinbergen, and the founding of ethology.* Chicago and London: The University of Chicago Press.

Burrow, M. (2013). The imperial Souvenir: Things and Masculinities in H. Rider Haggard's King Solomon's Mines and Allan Quatermain. *Journal of Victorian Culture, 18*(1), 72–92.

Davies, G. (2000). Science, observation and entertainment: Competing visions of postwar British natural history television, 1946–1967. *Ecumene, 7*(4), 432–459.

Denis, A. (1966). *On safari.* London: Fontana Books (first published 1963).

Fyfe, A., & Lightman, B. (Eds.). (2007). *Science in the marketplace: Nineteenth-century sites and experiences.* Chicago: University of Chicago Press.

Gouyon, J.-B. (2011). The BBC natural history unit: Instituting natural history film-making in Britain. *History of Science, 49*(4), 425–451.

Guida, M. (2018). Ludwig Koch's birdsong on wartime BBC radio: Knowledge, citizenship and solace. In F. James, R. Bud, M. Shiach, & P. Greenhalgh (Eds.), *Being modern: The cultural impact of science in the early twentieth century* (pp. 293–310). London: UCL Press.

Hawkins, D. (Ed.). (1957). *The BBC naturalist.* London: Rathbone Books.

Huxley, J. (Ed.). (1961). *The humanist frame.* London: Allen & Unwin.

Lewis, J. (Ed.). (2007). *Starlight days: The memoirs of Cecil Madden.* London: Trevor Square Publications.

Long, P. O. (2002). Objects of art/Objects of nature. In H. Smith & P. Findlen (Eds.), *Merchants and marvels* (pp. 63–82). New York, London: Routledge.

McLuhan, M. (1964). The medium is the message. In M. McLuhan (Ed.), *Understanding media: The extensions of man* (pp. 23–35). New York: Signet.

Mitman, G. (1999). *Reel nature: America's romance with wildlife on film.* Cambridge, MA: Harvard University Press.

Rider Haggard, H. (1885). *King Solomon's mines.* London: Cassell.

Ritvo, H. (1987). *The animal estate: The English and other creatures in the Victorian age.* Cambridge, MA: Harvard University Press.

Sandbrook, D. (2005). *Never had it so good: A history of Britain from Suez to the Beatles.* London: Little Brown.

Scott, J. P. (1950). Methodology and techniques for the study of animal societies. *Annals of the New York Academy of Sciences, 51,* 1001–1122.

Scott, P. (1966). *The eye of the wind.* London: Hodder and Stoughton.

Shackleton, K. (1989, August 31). Great gifts and testing inheritance. *The Guardian*, p. 39.

Shapin, S. (1988). The house of experiment in seventeenth-century England. *Isis, 79*(3), 373–404.

Shapin, S. (1994). *A social history of truth*. Chicago, London: The University of Chicago Press.

Shapin, S., & Schaffer, S. (1985). *Leviathan and the air-pump*. Princeton: Princeton University Press.

Sielmann, H. (1959). *My year with the woodpeckers*. London: Barrie and Rockliff.

Swallow, N. (1966). *Factual television*. New York: Hasting House.

Yeates, G. K. (1957). The bird-photographer. In D. Hawkins (Ed.), *The BBC naturalist* (pp. 17–20). London: Rathbone Books.

3

David Attenborough: The Early Years—Fashioning Zoological Expertise On-Screen

Whilst Peter Scott had brought the tradition and values of amateur natural history on television in his series *Look,* and Armand and Michaela Denis performed big game cinematography with *Filming Wild Animals,* yet another idea for an animal programme had landed on the desk of Controller of BBC Television Service Cecil McGivern.

> From time to time the Zoological Society finances expeditions to bring back rare animals for the Society's collection. By co-operating on the film side we could secure some original programmes—and perhaps some prestige—in the joint organisation of an expedition.

> We know that animals provide us with first-class material, and the adventures of an animal hunter in Africa would provide us with material quite as legitimate and no more expensive than Mayhew's current affairs projects.[1]

[1] Head of Talks, Television, 'Hunting for rare animals: Cooperation with the zoological society', memo to Controller Programmes, Television, 17 July 1953, BBCWAC T6/444/1.

© The Author(s) 2019
J.-B. Gouyon, *BBC Wildlife Documentaries in the Age of Attenborough,*
Palgrave Studies in Science and Popular Culture,
https://doi.org/10.1007/978-3-030-19982-1_3

Mary Adams (1898–1984), the author of this memo, was the head of the Talks Department for BBC Television. Following McGivern's positive response—'I am *most* interested in the scheme and we must try to make it succeed'[2]—she asked David Attenborough (1926–) to further investigate the possibility of a collaboration between the BBC and the ZSL. This programme idea would lead to the series *Zoo Quest*, depicting Attenborough's adventures in remote places, collecting animals for the ZSL and visiting exotic people. The first *Zoo Quest* series was broadcast over six weeks between the end of December 1954 and late January 1955. Its success with audiences and the BBC management was immediate.

In retrospective accounts, David Attenborough insists that the originality of the *Zoo Quest* series rested on the fact that the episodes combined different features of a variety of existing animal programmes. Each episode had two components. Films shot in the field were shown with a live scripted commentary from Attenborough, narrating the expedition and developing the storyline. This footage showed animal collectors in action. It also provided a sense of the animals' natural habitat, including the human populations with which they cohabited. In between these segments were live studio sequences during which Attenborough and a guest from the London Zoo presented some of the animals mentioned in the narration and brought back to London. The first *Zoo Quest* series could be seen as a combination of the traditional studio-based zoological programmes of the kind presented by George Cansdale in the late 1940s and early 1950s, with the travelogue ones, of which the Denises' programmes were about to become the epitome (Chapter 2).

Another way of looking at these programmes is to say that their success came in part from their staging the BBC as collaborating with an institution, the ZSL, unambiguously associated with the imperial project, at a time when the British Empire was unravelling. As we will see in this chapter and the next, these series captured the last glow of the empire, turning it *in extremis* into televisual spectacle. In addition, the series' concept of a collaboration between the BBC and the ZSL for the production of natural knowledge shows that Attenborough's televisual imagination

---

[2]Controller Programmes, Television, 'Programmes in cooperation with zoological society', memo to Head of Talks, Television, 10 August 1953, BBCWAC T6/444/1.

went beyond bringing together existing programme formats. Rather, it demonstrates that quite early in his career Attenborough's reflection, when making programmes, was not just on the level of the programmes' content but also on the institutional level. The *Zoo Quest* series does not simply tell stories about the collection of animals in exotic places. It also conveys information about the institution producing them, the BBC, staging that institution as the partner of a learned society in an enterprise pursuing the production of knowledge. This take on wildlife programme making ascribed a key role to wildlife television in the public construction of BBC Television as a trustworthy source of facts about the world. The next chapter will take a closer look at the *Zoo Quest* series' contribution to relocating the British imperial project in wildlife television. But first, the present chapter examines the genesis of the *Zoo Quest* series and how it contributed to shaping wildlife television.

Wildlife Television as a Visual Performance

David Attenborough began work at the BBC in the Talks Department in September 1952. After reading Geology and Zoology at Cambridge, he graduated in Natural Sciences in 1945, and then spent some time working for a publishing company (Attenborough 2010). Mary Adams, the head of the Talks Department, who hired him, was a formidable pioneering figure who had been instrumental in shaping science broadcasting first on radio and then on television before World War Two (Jones 2012). A Cambridge-trained biologist, with several years of experience as a research scientist, she was a close friend of Julian Huxley's. Like most British left-wing intellectuals in the 1930s, Adams was a convinced eugenicist. She entered the BBC after giving a series of noted radio talks on related topics between 1927 and 1930. In 1930, she became one of the first radio producers specialising in scientific subjects. During her tenure, she developed a style of science broadcasting built on close collaborations between scientists and broadcasters, using science as a source of content to develop compelling radio programmes rather than using radio to popularise science. Above all else, Adams valued imaginative and entertaining styles of presenting. She did not conceive of her programmes as means of imparting facts and abstract ideas. Instead, she produced human interest stories explaining how scientists work. In her view, such a focus on human action was the most efficient way of getting audiences to listen to scientific content on the radio. When she invited others to produce science radio shows, she favoured speakers who demonstrated this ability to win audiences through the development of attractive storylines (Jones 2012). Adams remained a radio producer until 1936. That year, she became the first head of BBC Television's newly formed Talks Department and brought to the new medium the approach to science programme making she had developed for radio.

In 1949, her boss, Cecil McGivern, the controller of BBC Television's programmes, described Adams as 'a woman of considerable intellect and considerable emotion'.[3] He praised her intelligence, enthusiasm, hard work, ideas and persistence but also lamented her lack of method and her weakness in management and administration (Murphy 2016). However, internal correspondence at the BBC reveals that in 1952 the Talks Department was cruelly understaffed, and under-equipped.[4] In the early 1950s, television was a work in progress and, within the BBC itself, still needed to demonstrate its worthiness. In September 1952, Adams explained to McGivern:

> Producer manpower also provides some headaches. Peter Thompson will have left us; Andrew Miller Jones will be in America for part of the time; Grace Wyndham Goldie will carry a heavy schedule; Paul Johnstone will not be fully experienced. I have provisionally suggested using Gil Calder (as Director) to fill the Peter Thompson vacancy so that we can discover whether he suits Talks productions. Two trainees (Peacock and Attenborough) will have emerged from the school by February / March.[5]

In September 1952, two new recruits had just joined the Talks Department, both with a brilliant career ahead of them. Ian Michael Peacock, specialising in current affairs programming, would become the first head

---

[3]Cecil McGivern, 10 March 1949, Mary Adams, Annual Confidential Report, BBCWAC L2/5/1.
[4]Mary Adams (Head of Television Talks) to Cecil McGivern (Controller, Television Programmes), 24 June 1952, 'Revised talks schedule: October/November/December 1952', BBCWAC T32/330/3.
[5]Mary Adams (Head of Television Talks) to Cecil McGivern (Controller, Television Programmes), 18 September 1952, 'Talks schedules: January/February/March 1953', BBCWAC T32/330/3.

of BBC2 in 1964, before heading BBC1 in 1965, eventually leaving the BBC for an executive position with Warner Bros. in the early 1970s. He would play a pivotal role in developing the international co-production deal between the BBC and Warner Bros. which led to the production of *Life on Earth* (BBC 1979). David Attenborough, for his part, specialised early on in natural history programme making, and then took up management positions at the BBC. He would succeed Peacock as head of BBC2 in 1965 before becoming the director of television programmes in 1969. He quit in 1972 and returned to programme making, working notably on *Life on Earth*.

Upon entering the BBC, newcomers received training in television making. They first had to take a general course for a month. This alternated lectures about the BBC, or more general topics such as 'The Writer and Television' (a lecture delivered by C. P. Snow) or 'The Documentary Film in Relation to Television' (a lecture delivered by John Grierson), with practical activities such as shadowing a producer or watching and discussing programmes.[6] Then, trainees served as studio managers for two months, after which they were attached to a couple of producers for a few weeks. Attenborough got to work with John Read and George Noordhof. Both specialised in science programmes, and they were the successive producers of the main animal programme in the early 1950s: George Cansdale's *Looking at Animals* (1951–1954). Eventually, having trained for five months, the trainees could start producing small-scale programmes on their own, before moving on to longer formats. After completing his training, Attenborough was tasked with producing a number of short programmes, ranging from gardening to current affairs. Some of them were significant to cement his association with natural history content and to enable him to develop a network within the milieu of natural history and zoology, starting with Julian Huxley, a key figure of the popularisation of the life sciences in the immediate post-war years.

Attenborough's first assignment as a producer was a twenty-minute programme broadcast on 13 January 1953, under the title *The Coelacanth*. It provides us with insights about how he approached television production

---

[6]Television Circular No. 281A, 1st April 1951. Television Training Scheme. Trainees from outside the Corporation, BBCWAC T16/240/1—Staff Training Department. Television training course, 22 September to 17 October 1952, BBCWAC T16/240/1.

when his career was just starting. The programme was topical, following the capture of one specimen of the prehistoric fish off the coast of Madagascar over Christmas 1952. This was an important event for scientists studying evolution. A rare occurrence which offered the hope of shedding some light on the process through which living organisms transitioned from water to land. Invited to talk about the discovery was Julian Huxley, a recognised authority among television viewers on all matters related to natural history and evolution. In haste, Attenborough wrote a script for Huxley, letting him tell the story and explain the significance of the event. Huxley received it on the day before the programme so he could rehearse.[7] The programme, transmitted live from Alexandra Palace, Studio A, with four cameras, presented visually varied content.

In 1952, as most television broadcasters were coming from radio, the dominant professional culture at the BBC was that of radio broadcasting. In a note from June that year, Cecil McGivern, the head of BBC Television, stressed that far too much emphasis was being put on the spoken word in television output at the expense of 'the thing seen'. To counter what he perceived as an alarming influence of radio on television, McGivern demanded that more efforts be put into developing the visual aspects of programme making.[8] This was a request that Attenborough was to follow in the first programme he produced on his own. The most radio-like sequences of Huxley directly addressing the camera alternated with others, much more visual, which showed the story rather than telling it, using displays of fossils and models brought from the Natural History Museum, diagrams positioning the coelacanth in its evolutionary context, and two silent films (specially shot by Norman McQueen)—one of a lungfish and the other of pompanos—to highlight, through comparison, the morphological characteristics of the lungfish. These visual representations enabled viewers to grasp for themselves the significance of the coelacanth. As a whole, the programme was in accordance with Adams's preferred style of science broadcasting. It privileged narration by a guest scientist, thereby offering an informed personal point of view, making it easier for viewers

[7] Secretary to Huxley, 12 January 1953, BBC WAC TVART 1—Sir Julian Huxley—Talks File 1.
[8] Controller, Television Programmes to Head of Television talks, 'Talks Schedule: October/November/December', 16 June 1952, BBC WAC T32/330/3.

to relate to the content. In line with McGivern's instructions to producers, it also emphasised television's capability to show things.

Animal Patterns: Framing Television Watching as an Informal Learning Experience

From Attenborough's close collaboration with Huxley when preparing and delivering this programme sprung an enduring relationship which lasted beyond Huxley's involvement with the BBC until his death in 1975. Following the transmission of the programme, the two men began exchanging books, and Attenborough sought Huxley's advice before answering viewers' letters. These sustained exchanges led Huxley to form an image of Attenborough as someone with whom he could collaborate to bring more zoological content to television. Attenborough, for his part, was eager to develop as a natural history producer and suggested programme ideas to Huxley, who then pitched them to Mary Adams. In March 1953, Attenborough mentioned to Huxley the idea of a series of programmes on the topic of animal coloration. In his reply, Huxley praised the idea—'I think your suggestion an admirable one, and it stimulated me to produce this program'—before advising his younger correspondent to read Niko Tinbergen's *The Study of Instinct* (1951) and Cambridge zoologist Hugh Cott's textbook *Adaptative Coloration in Animals* (1940).[9] Writing on the same day to Mary Adams, Huxley offered the idea of 'a possible 4 "talks" (would you call them T.V. Talks?!) on the "scientific natural history of colours and patterns"'. The series, he insisted, 'would make an excellent TV program—exceedingly visual, with the scientific interest *and* that of natural history'.[10] Huxley also mentioned that the idea originally came from Attenborough, adding: 'I should imagine he would be quite competent to collect the material.' Making plans for the programme, Huxley

---

[9]Julian Huxley to David Attenborough, 9 March 1953, p. 1. BBCWAC TVART1—Sir Julian Huxley—Talks File 1.

[10]Julian Huxley to Mary Adams, 9 March 1953, p. 1. BBCWAC TVART1—Sir Julian Huxley—Talks File 1.

suggested to both Adams and Attenborough that 'Tinbergen should be brought in'.

Taking ownership of the idea, Attenborough began shaping it. In an early discussion with Mary Adams, he argued that three programmes would be enough. The fourth one, in which Huxley wanted to have Tinbergen collaborate actively, was 'rather more difficult, intellectually, than the other three and … strayed rather a long way from the general theme of animal colouration'. By contrast, Attenborough felt that 'the remaining three programmes cohere and might form a very interesting trio'.[11] An initial thought had been that these programmes should be part of another series. 'Could it be fitted in as part of a more general scientific series [?]" asked Huxley.[12] Attenborough argued against it: 'To try and incorporate them in the general "British Countryside" series would, I think, be a pity.'[13] This would lead to the exclusion of the non-British examples— 'and most of the most spectacular are, of course, non-British'. Further, this would mean having only one episode on colouration 'in order not to over balance the series—which would be a waste of a lot of very good material'. Another consequence, which Attenborough did not discuss, was that audiences were more likely to take notice of a stand-alone series than of an episode in a more general series. Eventually, Adams convinced Huxley that three programmes would be enough and invited him to have a conversation with Attenborough. Based on Huxley's synopsis and the list of examples he provided, Attenborough set off to outline the series and started gathering material: film sequences, photographs, live animals, and museum specimens.

Attenborough conceived of the programme primarily as a visual performance, and at the pre-production stage, he gave special attention to the technical means that would serve to create the viewers' experience. The general idea was to have 'Huxley in the studio, in front of drapes, illustrating his points with either live animals, models, stuffed specimens,

---

[11] David Attenborough to Head of Talks, Television (Mary Adams), 10 March 1953, BBCWAC T32/330/4.

[12] Julian Huxley to Mary Adams, 9 March 1953, p. 1. BBCWAC TVART1—Sir Julian Huxley—Talks File 1.

[13] David Attenborough to Head of Talks, Television (Mary Adams), 10 March 1953, BBCWAC T32/330/4.

blow-up photographs, films or animations'.[14] The detailed planning of each episode, which would be transmitted live, left little room for the unexpected, such as animals escaping in the studio during transmission. And so, Attenborough also signalled that 'it may be necessary ... to have an assistant zoologist to help in demonstrating'. The first programme would deal with camouflage, the second one with warning signals, and the last one with courtship displays. Building on experience, Attenborough noted that 'the programme will stand or fall on the ability of the viewer to see detail in the close-ups of many of these animals. I gather from George Noordhof that "Looking at animals" suffered badly in this respect when the programme was transmitted from Studio "A."' For this reason, he requested that a newer studio at Lime Grove be used rather than one older one at Alexandra Palace.[15] Being in an up-to-date studio would ensure having 'all the latest technical devices at our command and show the small objects in the greatest possible detail'.[16] One of these was 'a special attachment on one of the cameras' previously used once in a programme on eye surgery.[17] Paying such attention to the tools of visualisation enabled Attenborough to emphasise the possibility television offered viewers to generate knowledge of the natural world for themselves, through direct observation of natural phenomena. This notion featured prominently in the advertisement for the series.

The article Attenborough drafted to present the series in the *Radio Times*, a piece which eventually appeared signed by Julian Huxley, announced that

the first programme will deal with the various fascinating methods used to render animals inconspicuous. Their concealment value is not always obvious at first sight, and often their true significance can only be appreciated by seeing the animals in their natural surroundings. For that reason we hope

---

[14]David Attenborough, 'Animal form and pattern', memo to Head to Talks Television, 21 April 1953, p. 1. BBCWAC TVART1—Sir Julian Huxley—Talks File 1.

[15]David Attenborough to Head of Talks, Television, 21 April 1953, BBCWAC TVART1—Sir Julian Huxley—Talks File 1.

[16]David Attenborough to Julian Huxley, 8 May 1953, BBCWAC TVART1—Sir Julian Huxley—Talks File 1.

[17]David Attenborough to Julian Huxley, 1 June 1953, BBCWAC TVART1—Sir Julian Huxley—Talks File 1.

to reconstruct in the studio a piece of a tropical swamp and a scene from a Pacific island. Viewers will then be able to judge for themselves the success of the camouflage devices we shall show.[18]

In this text, Attenborough makes a claim familiar to naturalists: true understanding of animals originates in considering them in their natural habitat as opposed to looking at captive or stuffed specimens. This rhetorical move then allows him to focus the attention on the construction work poured into making the programme, arguing that it will enable viewers to judge for themselves the validity of what is presented to them. The series, setting out to reveal 'the true significance' of that which is not immediately obvious, turns the television studio, equipped with special tools for visualisation, into a space where knowledge can be generated. In turn, viewers become participants in the knowledge production endeavour. *Animal Patterns* contributed to framing television watching as an informal learning experience.

March 1953 was a busy period for Attenborough, who was trying hard to come up with new programme ideas to establish his reputation within the BBC as a producer of successful natural history content. At the same time as he was preparing the *Animal Patterns* series, he also produced an outline for a much longer-term project titled *Traveller's Tales*. Based on travel accounts, this series was to offer as much opportunities for ethnographic programmes as for natural history ones. In this respect it fitted in the tradition of wildlife film-making. Early wildlife cinematographers, such as Cherry Kearton (Gouyon 2011), often did not distinguish local populations from the wild fauna they interacted with in exotic places. This trend was still alive in the 1950s, as films by the Denises exemplify. They depicted exotic charismatic megafauna as much as they showed exotic people, integrating the film-makers' interactions with both in the same storylines. The initial treatment for *Traveller's Tales* indicates nonetheless that Attenborough was anxious to position his own output as different from existing models of wildlife television. Mary Adams's insistence on the importance of storytelling to engage viewers was crucial here:

---

[18] *Radio Times*, 26 June 1953, p. 32. Illustrated with photographs of two species of stick insects.

These programmes should, I feel, be first and foremost informal, personal stories—not travelogues or lantern lectures. The telling of an incident which has some sort of plot, a beginning and an end, would not only hold viewer's attention more easily than something which opened with maps plus facts and figures, but would also be an easier task for an inexperienced speaker.[19]

Attenborough further insisted that guest speakers would be encouraged to use visual means to tell their story, through the display of films, photographs, and objects.

There should, however, be ample opportunity for making the programmes visually interesting with photographs or objects that speakers brought back from their travels; indeed there should be a lot of opportunity for 'The story of how I found this little idol started in ... etc.' type of story.[20]

In 1953, television still was a comparatively young medium striving to break free from radio. Just as the methods of broadcasting were in their infancy, so too was the grammar of television programmes, which still had to be written. Any new programme idea was an experiment and contributed to shaping both the medium's and audiences' expectations. Attenborough was a member of the first post-war generation of television producers with no prior experience of radio; he had free rein to define television broadcasting.

Zoo Quest to Sierra Leone: The Importance of Being a Good Performer

The *Traveller's Tales* suggestion got a positive reception from Mary Adams, and she urged Attenborough to talk to Cecil Madden, who would in October that year discover the Denises and invite them to present their expedition films on television.[21] However, nothing tangible seems to have

---

[19] David Attenborough, *Traveller's tales*, March 1953, BBCWAC T6/430/1.

[20] David Attenborough, *Traveller's tales*, March 1953, BBCWAC T6/430/1.

[21] Mary Adams to David Attenborough, in Assistant Controller of Programmes Television to Head of talks, television, 26 June 1953, BBCWAC T6/430/1.

come out of the meeting and eventually, Attenborough only produced one programme based on his initial concept. *The Dreaded Savage: Adventure in the Orinoco* featured adventurer Alain Gheerbrant (1920–2013), 'a Frenchman, a soldier, a poet, a publisher', as the *Radio Times* introduced him,[22] to present *Des hommes qu'on appelle sauvages* (1952), the film of his encounter with the Guaharibo tribe, shot during his 1948–1950 Orinoco-Amazon expedition. Although the *Traveller's Tales* concept did not become the series he had envisioned, Attenborough had been inspired by the travel films he watched when looking for material for the project, as well as by the production of the *Animal Patterns* series with Julian Huxley, and in July 1953 he struck gold.

The idea this time was for a co-operation between the BBC and the ZSL to film a zoological expedition collecting animals in the wild. Attenborough had had plenty of occasions to discuss mounting a joint BBC-ZSL collecting expedition with the man who would become his partner in this project, Jack Lester, the curator of reptiles at the London Zoo. Lester had appeared in the second episode of *Animal Patterns*, about warning signals. To illustrate Huxley's talk, and upon Attenborough's request, Lester participated in the episode in order to handle a live rattlesnake. Lester was also a regular participant in another BBC programme featuring animals from the London Zoo, *Looking at Animals* (1951–1954), produced at the time by George Noordhof, one of the two producers with whom Attenborough finished his training. The London Zoo had emerged from the Second World War with a depleted collection. Notably, the reptile collection was only a shadow of its pre-World War Two self, as all the venomous snakes had been euthanised to prevent their escaping in London in the event of the zoo being bombed (Zuckerman 1988). The reconstitution of this collection was, no doubt, a subject of preoccupation for the man in charge of it, Jack Lester.

Shortly after the broadcast of *Animal Patterns*, whose last episode was transmitted on 15 July 1953, Attenborough mentioned his conversations with Lester to Mary Adams who, on 17 July 1953, wrote to the controller of the BBC's television programmes, Cecil McGivern, to 'discuss with

---

[22]The Scanner, 'Television talk of the week, adventure in the Upper Orinoco'. *Radio Times*, 4 December 1953, p. 15.

[him] an idea which seems to have exciting possibilities'.[23] At this stage, Cecil Madden, McGivern's assistant, was kept out of the loop. He was only informed of the project a year later, on 13 August 1954,[24] barely a month before the expedition was set to leave. He voiced his disagreement, but to little effect: 'I protested when I heard of the Attenborough Zoo idea, which I heard too late, being excluded from the ahead planning'.[25] By then, the project had been approved at the highest level of management with the award of a D. G. Grant, from a budget administered by the Office of the Director General of the BBC, and earmarked to finance ambitious projects considered of special value to the Corporation.[26] In the meantime, having received the go-ahead from Adams and McGivern, Attenborough sat down with Lester at the London Zoo, and the pair produced a more detailed synopsis for *In Search of the Emerald Starling*. This Attenborough sent to Adams, insisting in the cover memorandum that the project was 'still very unofficial as far as the Zoo is concerned', but that both the director of the Zoo, and the secretary of the Zoological Society were enthusiastic about it.[27]

This first detailed treatment for what would eventually become *Zoo Quest to Sierra Leone* insists on the scientific nature of the enterprise. The programme, as Attenborough envisaged it, was to tell the story of Jack Lester from the London Zoo, sent on an expedition to collect 'rareties', animals never seen in Europe or rarely:

> Although these two birds will be the primary object of the expedition [the Emerald Starling and Picathartes], Lester also hopes to collect two other rareties [sic.]—the Super Sunbird, the most beautiful of all the Sunbirds which has only been seen in Europe once before, when two specimens Lester brought back in 1950 were on exhibition in the Zoo for 3 weeks; and the

---

[23] Mary Adams to Cecil McGivern, Hunting for rare animals: co-operation with the Zoological Society, 17 July 1953, BBCWAC T6/444/1.

[24] Assistant to Head of talks, television to Cecil Madden, 16 August 1954, BBCWAC T6/444/1.

[25] Cecil Madden to Cecil McGivern, 'Armand Denis and "Zoo Quest"', 31 December 1954, BBCWAC T6/112.

[26] D. G. Grant Programmes, 'meeting with T. O. Tel. and Mr. T. Sloan on August 18, 1954', BBCWAC T6/444/1.

[27] David Attenborough to Mary Adams, 'In search of the Emerald Starling', 31 July 1953, BBCWAC T6/444/1.

Olive Colobus, a very rare monkey. In addition, however, the expedition will collect whatever other interesting creatures it finds.[28]

The programme format tried out in both the coelacanth programme and the *Animal Patterns* series informed the architecture of the *Zoo Quest* programmes, alternating between studio sequences and films. The former were to enable viewers to get close-up views of the animals collected, the films showing audiences the captures taking place and what the animals' habitat look like. The series would reveal 'the many diverse methods employed by zoological collectors for capturing birds, monkeys, crocodiles, snakes, etc.', in addition, 'The grotesque plants of the forest and the primitive African tribes met by the expedition, would provide many subjects of the greatest visual interest for the film cameraman.' The insistence on the visual value of the programme, the storytelling—a quest—and the ambition to show scientists at work, makes this first *Zoo Quest* series an attempt at producing a programme exemplifying all the features regarded at the BBC in 1953 as characteristic of forward-looking science television programme making.

The expedition's logistics show how the remnants of the British Empire and its networks served as a substrate for early British wildlife television to develop. The team of four, Attenborough, cameraman Charles Lagus, Jack Lester and another keeper from the London Zoo, A. J. Woods, whose responsibility would be to keep alive all the animals collected during the trip, left London on 6 September 1954, from Brompton Air Station in Kensington (south-west London), boarding an Airwork Ltd.-Hunting-Clan Air Transport Ltd. flight. The company, Hunting-Clan, had approached the BBC after the press office had placed an article announcing the expedition in the *Daily Telegraph*.[29] Beyond their safari specialism, the prospectus for the company highlighted their long-standing association with the British Government as a troops carrier to West Africa. After three days of travel, the four men and their gear landed in Freetown. Lester's previous colonial experience—he had been a bank clerk in Sierra Leone before the Second World War (Oates 1999: 9)—had secured the assistance of the director of agriculture of the Colonial Office in Sierra Leone. During

---

[28] David Attenborough, 'In search of the Emerald Starling', 31 July 1953, BBCWAC T6/444/1.
[29] Hunting-Clan Air Transport Ltd. to BBC, 'Expedition to Sierra Leone', 9 August 1954, BBCWAC T6/444/1.

the three months that the expedition lasted, Attenborough and his team used government rest houses and received help with transportation.[30] As Attenborough reported from Sierra Leone to his head of department,

> The benefit of going with someone who knows the country and the people has been enormous. Wherever we have been, the official world has done everything it could to help us. Many of the people concerned are Lester's personal friends and without their advice and guidance and general lubrication of official wheels I fear we would have still been in Freetown.[31]

The four men returned to London at the end of November 1954. Their flight ended on an epic note, as one of the plane's engines stopped working when crossing the Channel.[32] The expedition had brought back enough film material for six half-hour programmes, and after intense weeks of editing, the first episode aired on 21 December 1954. Jack Lester was the presenter. He narrated the filmed sequences and provided a zoological commentary in the studio when the camera showed captured animals in close-up. Attenborough, for his part, was in the overhead gallery, producing the programme, directing cameramen and instructing the vision mixer.

Reactions at the BBC, following this first episode, were lukewarm. This was due to Lester's performance, judged weak. Leonard Miall (1914–2005), who by then had succeeded Mary Adams as head of Talks, noted that although an excellent zoologist, Lester turned out to be a poor television presenter. 'He got everything wrong on the live transmission, and that also made him tongue-tied' (Miall 1994: 181). Miall asked Attenborough to replace Lester as the presenter for the remainder of the series, with A. J. Wood, the expedition's other man from the London Zoo, helping with the zoological information.[33] As the phrasing of the audience research questionnaire used for the programme shows, presenters were expected to

---

[30] Richardson to Attenborough, 29 April 1954, BBCWAC T6/444/1.

[31] David Attenborough to Leonard Miall, n.d. BBCWAC T6/444/1.

[32] See Attenborough to George Lines, 24 November 1954, BBCWAC T6/444/1.

[33] See letter from D. E. Knight to Dr. Harrison Matthews, 'Zoo Quest', 5 January 1955, BBCWAC T6/444/1.

be 'appealing' and 'likeable'. To be sure, a good performance had to convey a sense of authoritativeness and expertise, but, first and foremost it needed to be attractive and pleasing. Broadcasters' evaluations of Lester's performance in the first episode of the *Zoo Quest* series confirm what reactions at the BBC to the Denises' appearances had already suggested (Chapter 2). Early broadcasters placed a premium on the programmes' entertainment value, questions of epistemic authority only coming second to the performance. After the last episode of *Zoo Quest to Sierra Leone* had aired, the Programme Board praised 'an admirable series' and commended Attenborough for his 'commentary and handling of this unique material'.[34] Leonard Miall added his own congratulations:

> Your own appearance on the screen was not originally planned and decided upon after Lester's weakness as a performer in the first programme. In fact, your own performances have added enormously to the value of the series.[35]

Within the BBC, this first *Zoo Quest* series established Attenborough's status as a creative mind and a very talented on-screen performer. For his part, Attenborough was keen to define *Zoo Quest* to his colleagues as a successful technological innovation. A contentious issue when planning the series had been the type of film to use in the field. The choice was between 35 mm and 16 mm. In terms of filming in the field, the main difference between the two gauges was not so much the camera's weight as the bulkiness of the film stock: 35 mm reels are much larger and heavier than 16 mm ones. This pleaded in favour of the latter.[36] However, at the time, 1953–1954, there were no 16 mm editing facilities at the BBC. The people managing film stock and equipment at the Corporation therefore strongly opposed the idea of using 16 mm film as they estimated the cost of outsourcing the editing of 16 mm film at twice that of in-house 35 mm editing.[37] Besides, the BBC film department only stocked 35 mm

---

[34]Head of Talks, Television, 'Zoo Quest', memo to David Attenborough, 19 January 1955, BBCWAC T6/444/1.

[35]Head of Talks, Television, 'Zoo Quest', memo to David Attenborough, 19 January 1955, BBCWAC T6/444/1.

[36]Talks Organiser, Television, 'Sierra Leone expedition', memo to Programme Organiser, Television, 18 May 1954, BBCWAC T6/444/1.

[37]Talks Organiser, Television, 'Expedition to Sierra Leone', memo to Programme Organiser, Television, 22 April 1954, BBCWAC T6/444/1.

cameras, which meant that a 16 mm one had to be found. Eventually, Lagus brought his own camera, and the decision to equip the BBC with 16 mm editing facilities tilted the balance in favour of using this film gauge for *Zoo Quest*. Another debate had been about whether Charles Lagus should shoot in colour or black and white. Cyril Jackson, Talks organiser, had come up with the suggestion that colour could be used, anticipating the time 'when colour television opens'.[38] After several weeks of discussions, the head of BBC Television, Cecil McGivern, went for colour. Less than a week after the last episode had aired, Attenborough authored a two-page memorandum, presenting the series as evidence that using 16 mm colour film for television programmes, far from problematic, was in fact revolutionary.

Writing to Leonard Miall, with a copy to the supervising film editor, Attenborough methodically countered the various arguments which had been voiced against 16 mm with evidence that they had been unfounded. Despite predictions that filming in 16 mm would result in a grainy image on transmission, 'many people technically knowledgeable have told me that they were quite unable to say whether the entire film was 16 mm or 35 mm'.[39] Editing had not been as problematic as prophesied; using 16 mm film made it even easier: 'The disadvantage of not being able to view satisfactorily in the hand was balanced by the advantage of having so much less celluloid to deal with.' The simplicity of the equipment, an Acmiola and a projector, made editing quicker and more straightforward. Notably, 'The projector has the additional advantage of making it possible to view the cutting copies on the cutting room wall, instead of having to struggle for theatre bookings.' With this first *Zoo Quest* series, Attenborough could claim to have successfully shaken the established way of handling film at the BBC. Up against a culture inherited from cinema, the first *Zoo Quest* series stood as evidence that a television-specific alternative was possible.

---

[38]Talks Organiser, Television, 'Sierra Leone expedition', memo to Programme Organiser, Television, 18 May 1954, BBCWAC T6/444/1.
[39]David Attenborough, 'Use of 16 mm. film in Zoo Quest', memo to Head of Talks, 27 January 1955, BBCWAC T6/444/1.

I was interested to get a telephone call from a commercial agent asking me incredulously if all the Zoo Quest film *was* on 16 mm. When I told him it was so, he said that it had caused a great deal of interest in the world of commercial television and that Harry Alan Towers was of the opinion that if it were true that we had used 16 mm. on these programmes it would revolutionise his approach to the whole business of television films.[40]

Commentators stressed the collaborative nature of the series, between the broadcasting institution and a learned society, the ZSL. It cast the BBC as a public body able to participate in scientific enterprises and in this way increase the common stock of knowledge. Leonard Harrison-Matthews, the scientific director of the London Zoo, authored an article in *The Times* celebrating a successful joint venture. According to his account, the expedition had three goals of equal importance: to collect animals for the London Zoo, to document the biology of certain rare species, and to record on film animal life for showing on television. Declaring the expedition a success on all counts, Harrison-Matthews expressed his hope that similar ones to other parts of the world would soon be organised.[41] Media content such as Harrison-Matthews's piece in *The Times* contributed to publicly fashioning the BBC as a potential participant in the scientific exploration of the natural world on the grounds of the expertise it could deploy in relation to film-making and broadcasting.

Already, in 1955, Attenborough personified this collaborative approach. Soon after the last episode of the series had been broadcast, John James Yealland, the assistant editor of the ZSL's popular zoological magazine *Zoo Life*, invited him to author a piece narrating the expedition for the spring 1955 issue. The text appeared alongside other articles by specialists of the fauna and flora of West Africa, upholding the notion that such collaborative expeditions were valid means for the zoological exploration of the world. It also placed its author, Attenborough, on an equal footing with the other experts who contributed to the issue. The feature served as a space for him to tell the story of his own apprenticeship, thereby making

[40]David Attenborough, 'Use of 16 mm. film in Zoo Quest', memo to Head of Talks, 27 January 1955, BBCWAC T6/444/1.
[41]Leonard Harrison Matthews, 'Animal hunting in Africa', *The Times*, 4 January 1955, p. 7.

up for his inexperience and lack of expertise in animal collecting. Attenborough starts by admitting his ignorance and goes on to narrate how, as the expedition progressed, he became attuned to the environment, and the topic, and his senses were educated: 'My eyes became more skilled in distinguishing the inhabitants of the forest' (Attenborough 1955: 12). The text evolves into a narration in the first-person plural, abolishing the boundaries between zoologists and television crew. Be they from the London Zoo or the BBC, all expedition members become equal participants in the zoological adventure. The trajectory set with the first *Zoo Quest* series continued with the two following ones, to Guiana (1955) and Indonesia (1956), resulting in 1957 with Attenborough's public status having evolved from that of a famous TV producer to that of a trusted 'explorer and zoologist'.[42]

Zoo Quest to Guiana Firmly Establishes Wildlife Television as a Participant in Science

Zoo Quest to Sierra Leone had provided evidence that such a programme could be made; the next one, *Zoo Quest to Guiana*, converted the attempt. As soon as the last instalment had been broadcast, Attenborough, building on his growing reputation within the BBC, began planning his next expedition along the lines of the previous one. The same BBC film crew, Attenborough—as director/producer—and Lagus—as cameraman—were to follow two curators from the London Zoo as they collected animals in the South American country. Whereas planning the previous expedition had taken nearly a year, this one only took three months to organise, and the party was off in May 1955 for three months. Just like in the British colony of Sierra Leone, Attenborough and his companions benefited in Guiana from the help and support of the British colonial administration. The resulting series offered the same blend of wildlife and anthropological content as the previous one, likewise presented from the studio by Attenborough. It similarly drew much applause from television audiences and

---

[42]Audience Research Report, 'Zoo Quest for the Paradise Bird, 1—The Barrier Reef, produced by David Attenborough', 2 December 1957, BBCWAC R13/462/1.

BBC executives alike. However, the similarities in content and structure between these two first series should not obscure the fact that the latter was a turning point in the fashioning of Attenborough's public identity as a trustworthy source of knowledge of the natural world.

This is down, first, to Attenborough significantly stepping up his production of support material for the series. He authored four articles in *The Times* which appeared whilst the party was still the field. These articles provided readers with a narration of the expedition as a scientific one. Acting as a kind of trailer for the yet-to-come television programmes as well as a behind-the-scenes disclosure, these newspaper articles prepared audiences to receive the series in the way intended: as a record of a scientific expedition. They had been Attenborough's idea. In February 1955, he approached the broadsheet, thinking that 'it would add enormously to the prestige of the expedition and the resulting programmes if regular (say fortnightly) articles were published by *The Times* while we were in Guiana'.[43] The published pieces, illustrated with photos taken by Attenborough, bore titles such as 'Zoologists in Guiana. British Expedition Captures a Caiman', or 'Exploring the Rupuni. Zoologists Capture Monster Fresh-Water Fish'.[44] Narrated in the collective voice of the first-person plural, these texts cultivated the ambiguity between film-makers and zoologists, all subsumed into the category of zoologists. They framed the television programme as an account of zoological work and David Attenborough, its front man, as a zoologist.

Zoo Quest to Guiana forwarded the notion that the BBC was an institution able, through programme making, to participate in the generation of expert knowledge about the natural world. Throughout the series, as in the book-length account published afterward, Attenborough, just as he had done in his 1955 *Zoo Life* article, staged his apprenticeship, his transformation from a television producer into a zoologist. This was made both easier and necessary when, at an early stage of the expedition, Lester fell from a horse, breaking ribs. This unfortunate accident prevented him from further active participation, and his role became that of

---

[43] David Attenborough, 'Pre-publicity for Guiana Quest', memo to Head of Talks, Television, 1 March 1955, BBCWAC T32/1798/1.
[44] Attenborough, 1955, 'Zoologists in Guiana', *The Times*, 11 May 1955, p. 11; Attenborough, 1955, 'Exploring the Rupuni', *The Times*, 28 May 1955, p. 7.

a teacher instructing Attenborough, who was now in charge of catching the animals. From being a mere witness, Attenborough became an actor in the knowledge-production enterprise, which required him to receive 'an education in forest lore' (Attenborough 1958: 39). For example, in the opening episode of the series, a sequence depicts the capture of a sloth. It opens with Lester leading Attenborough toward a tree and pointing his finger at something up in the branches. As Attenborough recounted in his voice-over commentary:

> Then, suddenly, Jack spotted, almost hidden in the branches of a tree, a mysterious moving smooth shape. It was one of the most extraordinary and fantastic animals in the world: the sloth.[45]

Next, filmed climbing up the tree, Attenborough captures the sloth and passes it on to Lester, who bags it whilst Attenborough comments that the capture was an easy task. The book published shortly after the broadcast told similar stories and can be read as the equivalent of a *bildungsroman*, a coming-of-age story, the first half containing several accounts of the education of Attenborough's senses of sight and smell: "'Is there anything up there, or is it my imagination?" he said softly. I could see nothing. Jack explained more carefully where I should look and at last I saw what he had spotted' (Attenborough 1958: 84). Accounts by field scientists, such as primatologists, often include acknowledgement of their assistants' contribution, usually local inhabitants, who taught them how to move in the bush or how to identify animals (Rees 2006). Likewise, to ascribe to Lester the role of a teacher of local knowledge, placed the London Zoo curator in the position of a field assistant to Attenborough, who was seen performing the actual deed of collecting (Gouyon 2011). As Attenborough had already noted in his commentary for *Zoo Quest in Sierra Leone*,

> As I soon learned, you don't catch many animals by wandering aimlessly through the bush. The only really sensible way to set about the business is to consult the local African hunters. It is they who know where and when to look for what.[46]

---

[45] *Zoo Quest to Guiana* (BBC 1955), episode 1. Original transmission date: 13 September 1955.
[46] Attenborough [script for *Zoo Quest to Sierra Leone*], 'Recording I', n.d. BBCWAC T32/1798/1.

If the act of collecting animals is at the root of knowledge production, showing Attenborough fortuitously becoming responsible for it symbolically demotes Lester and places him in a subaltern position, conversely upgrading Attenborough's epistemic status. The fact that he had read Zoology and actually held a degree in Natural Sciences from the University of Cambridge eased his public transformation from a TV producer into a zoologist. Yet, his university degree is not mentioned in the public content related to *Zoo Quest*, and by his own admission, Attenborough had never practiced zoology after graduating. As far as television audiences were concerned, the source of this expert status was his performance in the *Zoo Quest* programmes. In turn, the staging of Attenborough progressively becoming a zoologist, characterised the BBC as an institution capable of constituting expertise in natural history and zoological matters.

Zoo Time: The ZSL Withdraws Its Support

A measure of the BBC's success with the fashioning of Attenborough as an expert television zoologist is the ZSL's intense defensive reaction against the Corporation. It took the form of a television unit installed in the London Zoo in 1956, whose existence effectively rendered impossible any further collaboration between the BBC and the ZSL. Even before *Zoo Quest to Guiana* had been entirely transmitted—two episodes remained to be broadcast—Solly Zuckerman, the ZSL's honorary secretary, announced the establishment, within the zoological garden, of 'a resident Unit to produce films and television programmes'.[47] Discussions to establish a film unit at the London Zoo had begun in earnest at the same time as the *Zoo Quest* expedition to British Guiana was on its way back. The end of the expedition had been hectic. A few days before the expedition was due to fly home, Jack Lester had shown the symptoms of a severe bacterial infection, an amoebiasis (which would eventually kill him), and had urgently been repatriated. To take charge of the animal collection and

---

[47]Solly Zuckerman and Leo Harrison Matthews, 'The Zoological Society and Television', press release, 6 October 1955, ZSL archives. Folder 'Establishment of Granada TV and Film Unit—1955–56.

organise its return to Britain, the ZSL had to send another curator in his stead, John James Yealland.[48]

These events were taking place in the context of the opening of British television broadcasting to competition (September 1955). This political decision had prompted Sidney Bernstein, chairman of Granada Television, to initiate a discussion with Solly Zuckerman on the possibility of a collaboration between Bernstein's newly formed television company and the ZSL. Bernstein had just signed with the Independent Television Authority (ITA) the contract establishing Granada Television as one of its subsidiaries for the North of England (Yorkshire and Lancashire). He was on the lookout for attractive programme content to fill in his schedule. A friend of Julian Huxley's, Bernstein had already been involved in the production and diffusion of films with the ZSL in the 1930s when Huxley was secretary of the ZSL. At the time, Bernstein owned a successful chain of cinemas in London and the rest of the country. A seasoned entrepreneur in the show business, Bernstein was well aware of the high entertainment value of animals in visual media.

Zuckerman's exchanges with Bernstein prompted him to engage, in July 1955, in a review of the London Zoo's television policy. In August, Bernstein wrote a formal, three-page application, detailing his plans for a TV and film unit based at the London Zoo.[49] To Zuckerman, Bernstein's proposal was 'tantamount to the Zoological Society itself, with [Granada's] help, retaining the exclusive rights of production' of film content about the Zoo.[50] The letter so inspired Zuckerman that he used entire paragraphs from it in his own argumentative papers making the case for a Granada-operated TV unit. The ZSL council was asked to validate his endeavour at its September meeting. Zuckerman's own preamble to these papers suggests a degree of acrimony toward the BBC, referring unambiguously to the expenses the Zoo had incurred to repatriate Lester from British Guiana

---

[48]The Zoological Society of London, ZSL archives, Council Monthly papers January 1954–December 1955. Eventually, Jack Lester died, in August 1956. As a compensation, the ZSL paid £2500 to his widow and contributed a £700 grant toward the education of her two younger twin daughters.

[49]Sidney Bernstein, 'The Zoological Society and Television', letter to Solly Zuckerman, 29 August 1955. ZSL archives, 'Establishment of Granada TV and Film Unit—1955–56'.

[50]Solly Zuckerman, letter to Sidney Bernstein, 1 September 1955. ZSL archives, 'Establishment of Granada TV and Film Unit—1955–56'.

and send a replacement, as well as to Attenborough appropriating the status symbols of zoological expertise through his television programmes:

> Television ... has provided us, as its popularity has grown, with an increasing amount of publicity and with a trivial amount of revenue. But we have paid for this not only with a disproportionate measure of trouble, but also with some embarrassment, due at least partly to the fact that some of the T.V. personalities involved not surprisingly found themselves becoming more important on the T.V. screen than the zoological topics they were expounding, and the animals they were using as illustration.[51]

From Zuckerman's perspective, the relationship between the Zoo and the BBC was an exploitative one which threatened to undermine the Zoo's cognitive authority and to turn it into a back garden for the BBC. The opening of television to competition was an opportunity for the ZSL to regain control over the presentation of zoological topics on television and restore its authority in the public eye. On 13 September 1955, having heard of Zuckerman' plans, Adams and Attenborough had a fruitless meeting with Zuckerman and Harrison-Matthews to try and at least preserve the London Zoo's participation in the *Zoo Quest* programmes.[52] On 6 October 1955, the ZSL released its communiqué announcing the creation of 'The Zoological Society and Granada Television Film and TV Unit'.[53] Under the agreement, Granada did not get exclusive access to the Zoo and the collections but became a gatekeeper: any other broadcaster wanting to film in the Zoo, otherwise than for news coverage, first had to obtain an authorisation from Granada. In the context of competition, Granada was unlikely to grant such access to the BBC.

The Zoo's TV unit started its activities in April 1956. Its head was ethologist Desmond Morris, fresh from his PhD with Niko Tinbergen at Oxford. Morris also fronted the unit's main output, the children's weekly

---

[51] Solly Zuckerman, September 1955, 'Memorandum on television', p. 1. ZSL Archives, TV and Film Unit.

[52] Zoological Society of London, 'Note of interview with Mrs. Mary Adams and Mr. David Attenborough of the BBC on 13th September 1955', ZSL archives, Establishment of Granada TV and Film Unit—1955-56.

[53] S. Zuckerman and L. Harrison Matthews, 'The Zoological Society and Television', 6 October 1955. ZSL archives, Establishment of Granada TV and Film Unit—1955-56.

half-hour programme *Zoo Time*, as part of the ZSL's effort to create a public zoologist figure of their own. Kept at the London Zoo until 1963, the unit's existence effectively prevented the BBC from accessing the Zoo for seven years. This move on the part of the ZSL was an implicit recognition that the television medium could establish expertise. Consequently, it had seemed crucial to the ZSL, as a learned society granting expert status, to retain jurisdiction over the televisual representation of the body of knowledge it purported to command, and to maintain its ability to adjudicate over who could call themselves an expert zoologist. In practice, the ZSL's venture into television production meant that David Attenborough and Charles Lagus went on their own for their next expedition, to Indonesia. But in retrospect, far from undermining Attenborough's fashioning as a trustworthy source of knowledge about the natural world, the ZSL's withdrawal from the *Zoo Quest* project offered him the space to further establish his credentials as an expert naturalist and zoologist in front of television audiences. It also helped define the BBC as an institution that could, on its own, enact expertise of the natural world without having to rely on the visible support of such a learned society as the ZSL.

References

Attenborough, D. (1955). Expedition to Sierra Leone. *Zoo Life, 10*(1), 11–20.
Attenborough, D. (1958). *Zoo Quest to British Guiana.* London: The reprint society (originally published 1956, London: Lutterworth Press).
Attenborough, D. (2010). *Life on air: Memoirs of a broadcaster.* London: Random House.
Gouyon, J. B. (2011). From Kearton to Attenborough: Fashioning the telenaturalist's identity. *History of Science, 49*(1), 25–60.
Jones, A. (2012). Mary Adams and the producer's role in early BBC science broadcasts. *Public Understanding of Science, 21*(8), 968–983.
Miall, L. (1994). *Inside the BBC.* London: Weidenfeld & Nicolson.
Murphy, K. (2016). *Behind the wireless: A history of early women at the BBC.* London: Palgrave Macmillan.
Oates, J. F. (1999). *Myth and reality in the rain forest: How conservation strategies are failing in West Africa.* London, Berkeley: University of California Press.

Rees, A. (2006). A place that answers questions: Primatological field sites and the making of authoritative observations. *Studies in History and Philosophy of Biological and Biomedical Sciences, 37*, 311–333.

Zuckerman, S. Z. B. (1988). *Monkeys, men, and missiles: An Autobiography, 1946–88.* London: HarperCollins.

4

Wildlife Television, Empathy and the End of the British Empire

Zoo Quest to Guiana had ended with a studio sequence featuring Attenborough in conversation with Leonard Harrison Matthews, the scientific director of the London Zoo. Attenborough explained that, of all the animals he'd encountered and collected both in Sierra Leone and British Guiana, he'd got a 'a very soft spot for one very particular one that we've got in Africa'. This animal, Attenborough continued, was one which many viewers had written to him about, asking if they could see it again on television. Following this slightly teasing introduction in the manner of a music hall performer announced before her number begins, a young female chimpanzee, Jane, was brought onto the studio table and greeted as a beloved pet by Attenborough. In his account of the first expedition, published in *Zoo Life*, Attenborough had told how a hunter had brought the young ape to his party, 'cooped up in a tiny crate, very frightened and very wild' (Attenborough 1955: 17). At first, the chimpanzee would do nothing but 'spit, scratch and bite'. But after three days of good care, bananas and sweetened milk, she had settled down. Like the other animals collected, Jane, the chimpanzee, went to the London Zoo and was brought to the BBC studio to appear alongside Attenborough.

© The Author(s) 2019
J.-B. Gouyon, *BBC Wildlife Documentaries in the Age of Attenborough*,
Palgrave Studies in Science and Popular Culture,
https://doi.org/10.1007/978-3-030-19982-1_4

On transmission, as soon as she saw her familiar friend David, […] she clung round his neck and refused to budge throughout the programme. The sight of this sweet little frightened chimpanzee hugging her handsome young protector caused the whole television audience to 'ooh' and 'ah'. (Miall 1994: 182)

This last episode of the second *Zoo Quest* series, Jane's second television appearance, offered a repeat of the performance. Attenborough patted her, fed her grapes and tried to tickle her to make her laugh. He then recalled how touched he had been when visiting her at the London Zoo. The keeper had let him in the cage, Attenborough had called her name, and she had come 'running into [his] arms'.[1]

To an extent, with this simian embrace, Attenborough inserted himself in the cultural space of wildlife film-making that derived from the culture of imperial big game hunting. It had been commonplace for celebrity imperial hunters in the nineteenth century, such as Frederick Selous (1851–1917), to be pictured with dead exotic animals—trophies. Early wildlife film-makers, such as Cherry Kearton (1871–1940), in their effort to supersede big game hunters, while appropriating their status symbols, refashioned the practice, making it a habit of being seen in public in the company of chimpanzees and other exotic pets (Gouyon 2011). To appear in an affectionate relationship with a wild animal demonstrated their intimate connection with the wild and supported their claims of being experts of the natural world. Performing what looked like an affectionate embrace with a chimpanzee, Attenborough positioned himself and his programmes in this tradition. But whereas his forerunners had only used these displays of affection to fashion their expert identities, Attenborough integrated them in his repertoire of means to obtain knowledge of the natural world. To perform this embrace with a chimpanzee under the approving gaze of the scientific director of the London Zoo was to have an outwardly empathetic mode of interaction with the wild validated as an appropriate source of zoological knowledge.

The introduction of empathy with wild animals as one of wildlife broadcasters' knowledge production skills can be interpreted in the light of two

---

[1] *Zoo Quest to Guiana* (BBC 1955), episode 6, transmission date: 18 October 1955.

features of the British context in 1955: the opening of television broadcasting to competition, and the end of the British Empire. On 22 September 1955, commercial television broadcasting began in Britain, ending the BBC's monopoly and prompting an evolution of its wildlife output. In contradistinction to the BBC, rightly perceived to address the ruling middle classes, ITV and its subsidiaries pitched their programmes to working class audiences, and from the outset, independent television met with its public. Audience research conducted in 1957 found that 72% of viewers with a choice between watching ITV and the BBC chose the former, working-class viewers predominantly watching ITV (Waymark 2005). This change in the media environment prompted an adaptive reaction on the part of the BBC, with wildlife television a key element in the Corporation's reply to the challenge of competition. As the most successful BBC programmes of their kind in the first half of the 1950s, David Attenborough's *Zoo Quest* series stood on the frontline.[2] Displays of empathy toward animals, rooted in emotions, were thought to be more likely to appeal to working class audiences than displays of scholarly authority.

Displays of empathy toward exotic animals in such travel-based wildlife programmes as the *Zoo Quest* series also contributed to symbolically relocating the imperial project in nature, via television. In the decade following World War Two, as the British Empire started to disintegrate, travel became fashionable for the middle classes in Britain, contributing to keeping the imperial dream alive (Teo 2001). For example, from the mid-1950s onward, the Thomas Cook Group took to organising tours along the routes of colonial exploration used by Victorian and Edwardian explorers. Likewise, children and young readers were exposed to this romanticising of travel through the British Empire, as British popular print media, especially those aimed at a younger readership, became replete with fictional depictions of explorers' expeditions to capture zoo animals (Castle 2001: 148–149). These stories portrayed the deployment of Western technologies—trucks, planes, car, cameras and sound recorders—in the underdeveloped context of 'backward' countries. By contrast they exalted

---

[2]C. McGivern, 'New Patterns in BBC Television Programmes', *Radio Times*, 16 September 1955, p. 15.

the technological advancement of British society at a time when worries about the country becoming a second-order power were spreading wide (Sandbrook 2005). Beyond providing reassurance on the state of the metropole, these narratives—be they on television or in children's books—also nurtured the illusion of the persistence of a successful colonial project, relocated in nature.

Before the appearance of commercial TV, support for the *Zoo Quest* project's claims to trustworthiness had largely come from its visible association with the ZSL. But the intrusion of ITV into the landscape of wildlife television, which at first took the shape of the establishment of the Granada TV unit at the London Zoo, cut that source of cognitive legitimacy. And so, although David Attenborough did not repudiate the zoological repertoire to assert the reliability of the third series, *Zoo Quest for a Dragon* (1956), he drew from the additional one of empathy with animals and stewardship of nature, previewed in the last episode of the second *Zoo Quest* series. Besides, these technologically mediated demonstrations of empathy toward wild animals provided a moral justification for the televised exotic encounters, casting them as benevolent. They rendered Westerners' technologically mediated agency in the wild, as portrayed in the series, immune to contestation. From 1956 onward, through, notably, the *Zoo Quest* series, but also through the many variations on the theme—such as the programmes staging Gerald Durrell (1925–1995) collecting animals for his zoo in Jersey—wildlife television contributed to relocating the imperial project to wildlife, declaring the fauna of a shrinking British Empire the preserve of Westerners.

Handle with Care: Bringing the Empire Back Home Alive

In the aftermath of the second *Zoo Quest* series, Attenborough began planning a third trip. At first, the plan was to go to China, in a quest for the giant panda. The ambition was to link the programme to the news story of the ZSL sending Père David's deer to China in a bid to reintroduce the species in its native habitat. However, China did not grant permission for travel in the country and the project was abandoned. Instead, Attenborough proposed an expedition to Indonesia. The quest, this time, would be for one of the giant monitor lizards, animals popularly known as dragons of Komodo found on the small Indonesian island of Komodo. Attenborough and his cameraman Charles Lagus left London for Singapore in early May 1956, hoping to travel south to Komodo, collecting and filming animals along the way. In 1956, Indonesia was an independent country; the BBC crew could not benefit from the assistance of the British imperial infrastructure as had been the case for the two previous expeditions. Indeed, as Attenborough wrote to Leonard Miall from Djakarta, on 9 May 1956: 'In spite of all our letters of assurance from the Indonesian Embassy in London, everyone here is being as difficult as possible.'[3] Upon arrival, their filming equipment was confiscated, and the two men waited three weeks for authorisation to travel around the country. Eventually, Lagus, Attenborough and their collection of animals returned to London early in September 1956, having spent four months away, often in taxing conditions, both men contracting various tropical diseases and suffering minor accidents. When the first episode of the series aired a month later, on 5 October 1956, it was introduced by a visibly emaciated but still cheerful Attenborough. Without an expert from the London Zoo in the field and then in the studio to bestow a visible expert endorsement on the endeavour, Attenborough became responsible for establishing the programme's trustworthiness. Moving away from displays of manliness and the performance of traditional field skills as seen in the previous series, storylines for this new series brought forth an emotion-driven approach to nature, through the presentation of animals as individual characters.

Despite Attenborough and Lagus being on their own in their quest for a dragon, the concept of the series ostensibly remained the same. The expedition's objective still was to collect and bring exotic animals back to London to be displayed in the studio afterward and seen in the flesh at the London Zoo. However, captures of the kind performed in the previous two series are rare in the third one. When Attenborough appears on-screen, catching an animal, the scene often takes on a slightly comedic slant as a kind of humorous commentary on the displays of field skills and manliness

---

[3] Attenborough, personal letter to Leonard Miall, 9 May 1956, BBCWAC T6/439/1.

featured in the two earlier series. An example is the capture of a gavial, in the second episode. Attenborough is first shown carefully approaching a riverbank in a rather exaggerated manner. The film then cuts to a close-up shot of the animal, lying, unsuspicious, in the water, with nothing in the frame to provide viewers with a sense of scale. The voice-over commentary details the fact that this kind of crocodile is very dangerous, a man-eater which can reach a length of six meters. In the next frame, Attenborough carefully takes his shirt off before throwing it onto the animal. He then bends down to grab the beast, lifting up what is revealed to be a very small, young specimen. The sequence was consciously filmed and edited to elicit laughter, as Attenborough recalled in a later interview:

> So I had the idea that we would make a kind of joke of it. And that we would film it all in close-up and then I'd film myself taking off my shirt, and we hoped the audience would say, "He's not going to tackle that huge thing, is he?!" And only when I jumped on it would the people realise that it was just a tiny thing. We shot it that way and edited it that way.[4]

The capture of a python in the third episode receives a similar comical treatment. This time, Attenborough is first seen climbing up the tree where he's just spotted the snake, chopping down with a machete the branch on which it slithers. A short comedic sequence follows during which Attenborough repeatedly attempts to throw a bag on the snake's head, with the animal each time managing to escape. In this sequence, shot mute, Attenborough's mimics and demeanour are reminiscent of Charlie Chaplin, and meaning is conveyed visually, without having to rely on the commentary superimposed later in the studio.

These self-deprecating performances of animal capture can be interpreted as Attenborough deliberately trying not to appear as an expert animal collector—or at least as an imperfect one—to strengthen his status as one who can be trusted to dispense the truth about nature. Self-deprecation in relation to claims to knowledge has been identified as part of the cultural repertoire mobilised by individuals wishing to be perceived as

---

[4] 'David Attenborough's Zoo Quest in Colour' (BBC 2016).

truth tellers since the development of experimental science in seventeenth-century Britain. In this context, self-deprecation signals modesty. As the historian of science Steven Shapin notes: 'A man whose narratives could be credited as mirrors of reality was a "modest man"; his reports should make that modesty visible' (Shapin 2010: 101). The comedy-like capture scenes in *Zoo Quest for a Dragon* are filmed reports making Attenborough's modesty visible. A direct address to viewers, they suggest the presenter's self-confidence in the robustness of his trustworthy standing with audiences to the extent that he can visibly dispense with the status symbols of a zoo collector, asking viewers, instead, to trust him on his own terms. Attenborough's performance of a lack of expertise in traditional methods of zoo collecting opens up a space for him to carve his own form of expertise of the natural world, which does not rest on force and brutal confrontation, but instead, on care and affectionate attention to individual animals.

To perform his empathetic relationship with nature, Attenborough relied heavily on storylines centred on individual 'animal stars'. Whilst in Indonesia, with his producer/film director hat on, he was already thinking along such lines. Writing to Leonard Miall from the field, he reported the capture of 'several nice animals all trained for studio appearances (i.e. they will bite me on sight)—two young orang-utans, a bear cub and a 10 ft python will I think be the stars'.[5] Back in London, pre-publicity for the series focused on these animal stars as objects of affectionate care. This stands in contrast to the pre-publicity for the second *Zoo Quest* series— the four reports published in *The Times* during the expedition—which focused on the project's zoological dimension (Chapter 3). The third *Zoo Quest* series was announced by a full page in the *Daily Mirror*. The text describing how Lagus had brought Benjamin, the bear, back to his home in London was illustrated with a large photograph of 'Mrs Bridgette Lagus', the cameraman's wife, bottle-feeding the bear cub on her lap alongside the couple's daughter.[6] The choice of outlet for this story, the *Daily Mirror*, which in the 1950s was the go-to newspaper for working class readers (Thomas 2007), suggests that the deployment of strategies of cognitive legitimization based on displays of affectionate care of animals as opposed

---

[5]David Attenborough, personal letter to Leonard Miall, 9 August 1956, BBCWAC T6/439/1.
[6]*Daily Mirror*, 11 September 1956, p. 11.

to displays of zoological expertise, were considered more likely to attract and win the trust of viewers from the working classes, an audience which the BBC was at pains to reach in the wake of the opening of television broadcasting to competition (Waymark 2005). The front cover of the *Radio Times* issue announcing the new series, featuring Attenborough himself nursing Benjamin, drew from similar registers associated with emotions and domesticity.[7] As a demonstration that this type of evidence was considered essential in shaping viewers' perception of the presenter and the series as reliable sources of knowledge of nature, Attenborough, in the billing he wrote for the series, directed viewers' attention to 'the delightful little Malayan bear we've christened Benjamin, and whom you can see on the *Radio Times* cover this week taking nourishment from a milk-bottle'.[8]

From the outset of the series, empathy with wild exotic animals was thus presented as the defining feature of Attenborough's behaviour in the field. This was forcefully brought forward by embedding in the narration a contrast between the presenter's benevolence toward wild animals and local populations' lack thereof. The first episode, set in Borneo, centred on the quest for orang-utans: 'We wanted to go to [Borneo], particularly to catch orang-utans', explained Attenborough in the opening sequence. Following a description of Attenborough's and Lagus's first filmed encounter with the local Dayak people, a sequence depicts their trek in the rain forest, mostly crossing 'deep pools' on 'slippery submerged logs'. In the forest, Attenborough first identifies on the ground the remains of durian fruit 'which [he] knew was the favourite food of the orang-utan'. The Dayak guide confirms Attenborough's suspicion that 'the way in which it had been chewed showed that it had been eaten by an orang-utan'. Then, above, fifty feet up in the canopy, 'we saw a nest'. A few minutes later, Attenborough continues, a crashing noise in the branches above attracts their attention to 'a great furry red form swaying in the trees'. Following an observational sequence depicting an orang-utan moving in the canopy, the image cuts back to the studio and, with a concerned look on his face,

---

[7] *Radio Times*, Vol. 132, issue 1716, cover.

[8] David Attenborough, 'Zoo Quest. In search of a Dragon', *Radio Times*, 28 September 1956, p. 45.

Attenborough recounts how his intervention saved an ape from a spiteful Dayak:

> A few minutes after that last shot was taken there was an explosion, and I looked around and I saw that one of the Dayaks, who had come with us during the afternoon was holding a smoking gun. He had tried to shoot that orang-utan. I'm very glad to say he'd missed it and I turned to tell him what I thought about it, and it seemed to me really almost murder. But he said, "Well there are many orang-utans here, they steal my bananas, they steal all my crops, they are pest, I must shoot them". And away he went after it. But at least our conversation delayed things a bit and I'm very happy to say that he never caught it.[9]

The trek in the forest looking for orang-utans took on different meanings, depending on the participants: to the Dayaks, the Westerners' zoological trip was a hunting expedition. Observing this difference enabled Attenborough to emphasise his own empathetic approach to animals, setting it up as the moral grounding of his understanding of wildlife, and the source of knowledge in the series.

The story, in the second half of the same episode, of the young orang-utan Charlie, another of the series' animal stars, reinforces this notion of Westerners' moral superiority over local exotic populations when it comes to relating to wildlife, further establishing it as evidence of Attenborough's expertise. Echoing the story of Jane, the chimpanzee, in the first series, Charlie's story is one of rescue. Attenborough bought the ape, he tells viewers, for a large share of the expedition's tobacco, from a local man who'd captured it after finding it 'raiding his plantations at the back of his hut'. The man had locked the ape in a tiny crate, where Attenborough found it to be 'very frightened', all biting and scratching. After he'd brought it on the boat where he kept his animal collection, Attenborough transferred Charlie to a larger cage, leaving him to settle down. There follows a sequence demonstrating how Attenborough 'slowly … managed to win [Charlie's] confidence. Soon, every time I passed his cage he'd stretch

---

[9] David Attenborough, *Zoo Quest for a Dragon*, episode 1—Borneo. First broadcast 5 October 1956 on BBC television. Available online at https://www.bbc.co.uk/iplayer/episode/p00dgmd0/ zoo-quest-zoo-quest-for-a-dragon-1-borneo-part-one. Last accessed 23 November 2018.

out his hand to attract my attention, in the hope that he'd get more food.' Illustrating this commentary is a close-up shot of Attenborough and the ape amicably holding hands. In this filmed account of the taming of the orang-utan, Attenborough is the only human seen physically interacting with the ape. The sequence concludes with the disclosure that the food Charlie liked most was 'an egg', upon which revelation the camera cuts back to the studio. There, Attenborough informs viewers, is Charlie, safe and sound, 'back in London'.

The studio's display of Charlie invited viewers to engage with the exotic animal in the way Attenborough prescribed. Accompanying the ape was Laurie Smith, head keeper of the monkey house at the London Zoo. In his conversation with Attenborough, Smith did not impart any zoological knowledge about orang-utans, but his comments on how well the ape had settled down in his new environment and on his character contributed to constructing Charlie as an individual personality to whom viewers could relate. Attenborough, for his part, demonstrated his affectionate closeness with Charlie, stroking him under the chin and giving him an egg to eat, replicating live the sequence just seen in the film. Following the broadcast, visitors flocked to the London Zoo, hoping to see Charlie, and they brought eggs in an attempt to re-enact the scene witnessed in the television programme. As Laurie Smith noted: 'Over the past week-end the constant enquiry was to see "Charlie", impossible for the time being of course, but many other persons were pointing to our "Alex" and were quite happy to think they had seen Charlie in the flesh.'[10]

The narrative constructed around Charlie set the tone for the series and revealed the new kind of evidence Attenborough brought forth to support his claim to trustworthiness: his ability to develop an intimate, emotion-driven, empathetic relationship with wild animals, and his capacity to make them part of the social body. The relationship of mutual acknowledgement he entered into with the various animal stars in the series publicly demonstrated his credibility as a producer of knowledge of wildlife. It also provided viewers with a model to intimately interact with the animals presented in the series and produce knowledge for themselves. Attenborough thus enrolled his audience in support of his claims to trustworthiness,

---

[10]Smith to Attenborough, 10 October 1956, BBCWAC Folder T6/439/1.

constructing the series as a shared endeavour of production of knowledge about exotic nature. But to bring home the wildlife of a vanishing empire and provide viewers with an empathy-based model to engage with it was also to align the presentation of exotic wildlife on television with the way Peter Scott's series *Look* presented British wildlife from Bristol (Chapter 2). Scott's programmes, featuring amateur naturalist cameramen, encouraged television viewers to imitate these guests—for instance, to try and film badgers or birds in their back garden or in neighbouring woods. *Look* was an invitation to partake in an understanding of wildlife as something to be appreciated aesthetically and investigated through the lens of a camera. *Zoo Quest for a Dragon* framed wildlife to audiences as something to be related to and known through empathy. In both cases, the television programmes purported to provide those watching them with a template to engage with wildlife on an individual basis.

With *Zoo Quest for a Dragon* the BBC stood as an independent centre able to embody expertise in natural history topics, with Attenborough as one of its home-grown experts. The last episode of the series reached a reaction index (RI), a percentage quality rating by viewers, of 91. This made it the most successful programme of its kind at the BBC, a status previously held by the last episode of the previous series. Interviewing viewers, the audience research department of the BBC found confirmation of Attenborough's 'popularity as a speaker, explorer and lover of animals'. Viewers emphatically celebrated

> his gift for giving an unassuming but always compelling account of his adventures. … We regard Attenborough as the finest type of young Englishman—unpretentious, humorous, resourceful and humane with animals. A grand boy! How well he tells his story too.[11]

These audience research interviews reveal that Attenborough succeeded in coming across as a modest, trustworthy truth teller, and viewers positively perceived his love of animals as a credible source of expertise. As far as the BBC was concerned, these findings characterised the presenter as one of the most successful programme makers on the Corporation's payroll, both

---

[11] Audience research Report, Zoo Quest, 6- 'Dragons for Komodo', 3 December 1956, BBCWAC T6/439/1.

as a producer and performer. This led to his appointment as the producer of a new series, *Travellers' Tales*. Presenting films documenting expeditions, it quickly became a fixture in the television schedule and helped shape the professional culture of television broadcasting which was developing during the late 1950s to early 1960s at the BBC.

Participating in Science with Film-Making, in *Travellers' Tales*

In the late 1950s, when the conception still prevailed that television was a live medium inherited from radio, *Travellers' Tales* participated in the emergence of a professional culture of television broadcasting centred on the use of film. This approach would become predominant in the following decade. The idea for *Travellers' Tales* originated in the initial programme proposal that Attenborough had put forward in 1953. Back then, only one programme had materialised (Chapter 3). But in the renewed context of competition, the BBC was looking for innovative content. This provided Grace Wyndham Goldie, who had been enthusiastic about the original idea three years earlier, with the opportunity to revive it as parts of the Talks Department's offer of new programmes. And because no one else in the Talks Department had 'a comparable knowledge of expeditions and of the use of 16 mm film', Attenborough was tasked with producing the series from March 1956 onward. This meant sourcing films, editing them for broadcasting, and introducing them from a studio with the film-maker providing some context.[12] This approach to producing *Travellers' Tales* posed a practical problem as Attenborough was away a good share of each year, filming the *Zoo Quest* expeditions. To be able to keep transmitting the series regularly, despite his long absences from London, he suggested that, except for the *Zoo Quest* series—when he would be back anyway—the introduction of each programme should be filmed and edited with the main film. Arguing his case with Leonard Miall, the head

---

[12] Head of Talks, Television, 'Travellers' Tales', memo to Head of Programme Planning Television, 12 March 1956, BBCWAC T6/430/1.

of Talks Department, Attenborough highlighted the greater rationalisation of the programme production stream this would allow: 'If it were decided to treat them in this way, it would be possible to begin to make a stock pile', which would save studio costs and making programme planning easier.[13] To stockpile enough programmes for weekly transmission over the course of three months, which meant in effect preparing thirteen programmes, Attenborough turned the occasional facilities used for his annual *Zoo Quest* series—a room equipped with 'a Steenbeck editing machine, and an editor and an assistant editor'[14]—into a more permanent feature of the Talks Department's equipment, the cutting room becoming another key space for programme production alongside the studio control room. As such, *Travellers' Tales* encouraged a move away from live, studio-based broadcasting toward programmes entirely recorded on film.

This series helped cement the notion that the BBC could participate in the scientific exploration of the world, reinforcing the cognitive status of the *Zoo Quest* programmes and their presenter. In order to ensure a regular supply of film material for *Travellers' Tales*, the BBC contributed financially, or in kind, to scientific expeditions. This enabled the Corporation to secure the rights to broadcast these expeditions' film records. For example, the 1958 Oxford and Cambridge South American Expedition received £500, and Sir John Hunt's climbing expedition to the Caucasus, £150. In exchange, both ceded to the BBC the exclusive right to show their films. The broadcaster's contribution could also be in film stock, as was the case for Dr. Colin Rosser's and Professor Furer Hainendorf's expeditions to Nepal. In other instances, explorers simply agreed to go on an expedition to produce film material for the Corporation, thereby securing an outlet for their work. In the context of the competition with ITV, and of the partnership between Granada and the London Zoo (Chapter 3), these agreements enabled the BBC to stand as a patron of science and an institution able to embody expertise, or at the very least participate in the constitution of expertise. In turn, the introduction in the television schedule of a slot reserved for these films of scientific expeditions created a strand

---

[13]Attenborough, 'Travellers' Tales', memo to Assistant Head of Talks, Television, 19 April 1956, BBCWAC T6/430/1.

[14]Attenborough, 'Future Programmes', memo to Talks Organiser, Television, 8 February 1957, BBCWAC T6/430/1.

of programmes of which the *Zoo Quest* series could be a part. Proposing a schedule for the last quarter of 1956, Attenborough listed, under the same heading (1) the six episodes of his *Zoo Quest* to Indonesia; (2) three films by Norwegian zoologist Per Høst; (3) three instalments of the film report of the Oxford-Cambridge Far Eastern Expedition; (4) ethnographical films on Katmandu by Colin Rosser; (5) *the Peul Bororo people in West Africa* by Henry Brandt; and (6) *the Tuaregs*, a film bought in France from the Musée de l'homme.[15] Placing *Zoo Quest* in the same programming slot as other expedition films helped the BBC cultivate an audience for this type of programme content and associate Attenborough's name, as a television producer, with films of travel and exploration shot by scientists. This approach to scheduling implicitly bestowed on Attenborough the qualities and the cognitive authority of the zoologists, ethnographers, and other scientists responsible for the scientific films. But the integration of the *Zoo Quest* programmes in the *Travellers' Tales* series also reflected very positively on Attenborough, as a programme maker, within the BBC. A compilation of the audience appreciation indexes for the episodes broadcast in 1956 and 1957 demonstrated that the *Zoo Quest* programmes were amongst the most successful programmes of the *Travellers' Tales*.[16]

Travellers' Tales facilitated a rapprochement between the series produced in London and the wildlife programmes produced in Bristol at the NHU. At the end of 1957, Kenneth Adam, the controller of the BBC's television programmes, gave the newly established NHU in Bristol oversight of *Travellers' Tales*, which became a weekly programme spot receiving 'contributions from all sources in the exploration and natural history fields'.[17] The NHU was to make 'recommendations about the placing and order of different contributions in addition to being responsible … for production of some of the series'. This decision encapsulated in *Travellers' Tales* the notion that natural history and travel were tightly interwoven. It brought the *Zoo Quest* series, the other expedition films à la Attenborough, and

---

[15]Attenborough, 'Travellers' Tales', memo to Assistant Head of Talks, Television, 19 April 1956, BBCWAC T6/430/1.

[16]Head of Talks, Television, 'Travellers' Tales', memo to Controller of Programmes, Television, 24 October 1957, BBCWAC T6/430/1.

[17]Controller of Programmes, Television, 'Travellers' Tales', memo to All Members of Programme Board, 18 October 1957, BBCWAC T6/430/1.

the output from Bristol under the same natural history heading. From this point onward, *Travellers' Tales* stood on a par with *Look*, as Peter Scott's series was to run 'concurrently in another programme space'.[18] Adam's ruling made it a necessity for Attenborough and the people working in Bristol to collaborate. On 11 November 1957, Attenborough asked if Nicholas Crocker, the acting head of the NHU at the time, came to London often, and if so, whether they could meet: 'It would be nice to get to know one another.'[19] Then, from time to time, Attenborough pointed out film material he had received for *Travellers' Tales* but which could be of more interest to the NHU. Conversely, Crocker took to the habit of informing Attenborough of the content of programmes planned in Bristol to avoid overlaps and duplicates.

In 1959, BBC Audience Research found that travel programmes were the third most popular type of TV programme after news and plays. Part of this appeal could, just as it was with the *Zoo Quest* series, be related to cultural trends. The period witnessed a boom in overseas travel. By the end of the decade, more than one million Britons holidayed abroad every year (MacKenzie 2001). And as the British Empire was shrinking, travel TV programmes gave viewers the impression that far from retreating from the world stage, Britain, through the BBC—defined as a national service— was still reaching the far corners of Earth, projecting a reassuring image of confident imperialism. In this respect, Granada's wildlife television, limited to zoo-centred programmes, was weak. A confidential document compiled at Granada and comparing audience numbers between *Zoo Time* and competing BBC programmes, indicates that in the second quarter of 1958, Desmond Morris's weekly programme, live from the London Zoo, attracted on average 16% of the potential television viewing audience whilst the BBC's *Look*, over the same period, only gathered between 9 and 13% of this audience, depending on the time of broadcast. 'Only *Travellers' Tales*', the Granada document stated, 'gets higher ratings than *Zoo Time*.

---

[18]Controller of Programmes, Television, 'Travellers' Tales', memo to All Members of Programme Board, 18 October 1957, BBCWAC T6/430/1.

[19]David Attenborough 'Natural history programme plans', memo to Nicholas Crocker, 11 November 1957, BBCWAC T6/430/1.

… Its average rating is 19%'.[20] To attribute oversight of the *Travellers' Tales* series to the Bristol region enabled the NHU in Bristol to exploit Granada's weakness. Existing programmes administered in Bristol—Armand and Michaela Denis's *On Safari*, or films by ocean explorers Jacques Cousteau and Hans Hass—all with a strong escapist value, became episodes in the *Travellers' Tales* series. In addition, the NHU produced new programmes whose main appeal rested on their travel dimension.

Zoo Quest Becomes a Template for Wildlife Television

Direct or implicit references to the *Zoo Quest* concept abounded in the NHU's new output in the late 1950s to early 1960s. This is evident in the 1957 series *Faraway Look*. For this series, Charles Lagus accompanied Peter Scott as he went to Australia to preside over the international jury at the 1956 Olympic Games held in Melbourne. As Scott attended to his duties, Lagus filmed wildlife around Melbourne and in New South Wales. On their way back to Britain, the party visited New Guinea, New Zealand, Fiji and Hawaii. They returned in February 1957 with 30,000 feet of film and *Faraway Look* was broadcast in the following summer. Through his repeated association with Attenborough's series, Lagus had become a wildlife television personality of his own. As Scott and Lagus were departing for Australia, in October 1956, Attenborough penned a short piece in the *Radio Times*, titled 'Lagus of "Zoo Quest"'. In it he explained that although Lagus was hardly seen in any of the *Zoo Quest* films, 'his is the most arduous task, for while I am enjoying myself looking for animals, he may be lying on his back hugging the camera to his chest and trying to focus on a monkey high in the trees'.[21] To put Lagus and Peter Scott in the same programme was part of an effort to appropriate some of *Zoo Quest*'s appeal for Scott's programmes, and give them a kind of *Zoo Quest* twist. This short piece by Attenborough can be interpreted as his

---

[20]A. Anson to S.L.B. [Sidney Bernstein], 'Zoo and Animal Programmes Ratings', 6 August 1958, ZSL Archives, 'Granada Confidential File'.

[21]David Attenborough, 'Lagus of "Zoo Quest"', *Radio Times*, 19 October 1956, p. 45.

contribution to this effort. In the following years, the NHU went further in this direction and attempted to create a television personality who could deliver performances similar to Attenborough's. The ideal candidate would go on animal collecting expeditions and tell good stories about them but work for the NHU and not outside it. For a while, the broadcasters in Bristol found what they were looking for in Gerald Durrell (1925–1995). A popular writer, the author of *My family and Other Animals* (1956) had been known to BBC radio audiences since 1951 when he presented a series of radio talks about his experience as an animal collector.

Several of Durrell's radio talks had been produced by Eileen Molony (1914–1982), then the editor of the radio programme *Woman's Hour*. In 1959, Molony transferred to television, taking over the position of producer of *Look* and in this capacity had revived the *Faraway Look* series in 1959, sending Scott on a new tour abroad, accompanied by Tony Soper as a cameraman, this time to retrace Darwin's voyage to the Galapagos.[22] In the early 1960s, she became one of the champions at the NHU of David Attenborough's approach to wildlife television production, eventually extending him an invitation to host a programme produced in Bristol (Chapter 5). Molony's work with Gerald Durrell was part of the same effort. Durrell credited Molony with teaching him the art of performing to engage audiences (Durrell 1958).

In *My family and other animals*, Durrell recalls his childhood in Corfu as an origin story for his animal collecting passion. When the book came out in 1956, it landed on the desk of Tony Soper, who was then still producing *Look* in Bristol. Having read it, Soper spoke on the phone to Durrell and went to visit him and his wife in Bournemouth, where they lived, to discuss potential television programmes on the theme of animal collection. As it happened, Gerald and Jacqueline Durrell (1929–) had recently returned from a trip to Cameroon, which Gerald had documented on 16 mm film. Jacquie Durrell gives, in her 1967 *Beasts in My Bed*, a delightful account of their adventures, which led the Durrells to the palace of the Fon of Bafut. When Soper visited them, the couple was staying in a flat at the home of Gerald's sister, Margaret. They kept their collection of animals in the

---

[22] *Faraway Look* (1959–1960, BBC), first broadcast between 18 September 1959 and 6 November 1959.

garden and the garage, except for a chimpanzee named Cholmondeley, which stayed with the couple. Cholmondeley would soon join the cohort of wildlife television animal stars. Accompanying Soper to Bournemouth was Christopher Parsons, who had come to film some of the animals. The sequence later served as an introduction to the Durrells' first television series, *To Bafut for Beef* (1958), built around Gerald's 16 mm footage, and broadcast in April 1958 as part of the NHU contribution to *Travellers' Tales*. According to Parsons (1982: 118), Durrell's film coverage of the trip was not adequate, but Tony Soper deployed treasures of editing skills to construct a viable story.

The programme, in three parts, was an attempt at moving away from the *Look* template whilst retaining some formal aspects of Scott's series. The living room in the Durrells' flat was reconstructed in the studio, including their dog, Johnnie. As Jacquie Durrell remembered, the programme was 'deliberately off-beat in its presentation and it was a serious attempt on Tony's part to get away from the normal two people sitting in hard chairs, gazing inanely into a camera and indulging in a not very enterprising cross-talk act' (Durrell 1967: 100). In between clips from Cameroon, studio sequences featured Gerald and Jacquie Durrell's entertaining and humorous storytelling, as well as Cholmondeley's antics. Like the other adventure and travel-based programmes produced in Bristol at the time, *To Bafut for Beef* was an attempt to broaden the reach of the BBC's natural history output by distancing it from the relative dryness of its flagship series, *Look*. The retention of *Look*'s pointer to the domestic setting was a way to encourage viewers to relate to what was shown on the screen. At the same time, the style of presentation placed a premium on entertainment, as opposed to the restraint displayed in Scott's programmes. Part of the fun was due to the chimpanzee's stealing the show. The ape had been labelled 'a new TV personality' in the *Radio Times*.[23] Jacquie Durrell noted that 'our dear chimp was by now so full of himself and his own importance, having appeared on television many times in such august programmes as "Tonight", that he had become pompous and was determined to let no one else have a chance' (Durrell 1967: 100). Adding entertainment to the programme, the display of a tamed chimpanzee in this context was also

---

[23] *Radio Times*, 'Bafut Safari', 28 March 1958, p. 4.

a nod to the convention attached to wildlife film-making. Here again, the demonstrated ability to obtain the trust of a wild animal worked as evidence of Gerald Durrell's special access to the truth of nature.

To Bafut for Beef was well received and the NHU took an option on the television rights to the Durrells' next expedition. This was to be in Argentina and would offer six programmes, again to go into the *Travellers' Tales* slot. However, the BBC had not been able to commit a cameraman to accompany the Durrells in the field, and Gerald Durrell was not a film-maker. The Argentinian expedition eventually resulted in only one episode for *Look*, less dependent on film. Later, in 1962 Christopher Parsons would go with the Durrells in the field for the series *Two in the Bush* (1962), constructed using the *Zoo Quest* template. But in the interval, Gerald Durrell got busy developing his own zoo on the island of Jersey. Once the zoo was established, Durrell's 'effectiveness as a raconteur', his capacity for communicating with enthusiasm his passion for animals, 'his wonderful humour, descriptive powers and personal charm' (Parsons 1982: 119)—in other words his showmanship—were harnessed for another, studio-based series, *Zoo Packet* (1960), designed and produced by Eileen Molony. In a typical episode, organised around a scientific theme such as the anatomical adaptation of different species of monkeys to different environments, Durrell would narrate his animal collection adventures and display some inmates from his zoo. *Zoo Packet* was the NHU's direct response to Granada's *Zoo Time*. Durrell's programmes exemplify British wildlife television's general move, in the late 1950s and early 1960s, toward exhibiting the spectacle of exotic nature as opposed to British wildlife. By this time, *Look* was no longer focusing on British fauna. Peter Scott, who was about to launch the World Wildlife Fund (WWF) (in 1961), regularly introduced films depicting African fauna.

The *Zoo Quest* series and their posterity, as they developed in the second half of the 1950s, participated in suggesting a renewed relationship with nature to British audiences on the threshold of a post-colonial world. These programmes all strived for their success on the cultural climate of the perceived collapse of the British imperial project and the accompanying development of overseas travel. Although colonised people were achieving their independence, wildlife television flourished as a refashioning of the imperial project, through the colonisation of nature. This new form of colonisation rested on the deployment of a new kind of expertise based on displays of empathy for, and love of animals, suggesting that Westerners enjoyed an understanding of nature superior to that of native populations. The visible deployment of filming technologies in these countries, as a material means of mediating this new appropriation of natural resources, made it even more difficult for local populations to contest this new form of colonisation.

The NHU's oversight of *Travellers' Tales* only lasted a year, but this was enough to turn its output toward exotic wildlife. By the end of 1958, the series returned to the Talks Department in London, through the creation of the Travel and Exploration Unit, placed under the aegis of Attenborough, who became the series' editor. He completed two further *Zoo Quest* series,[24] and then, the focus of his own television work began shifting away from wildlife toward more geographical and anthropological topics. Pitching his next endeavour, *People of Paradise* (1960), Attenborough wrote:

> I would not suggest that the series should be concerned in any major way with natural history and indeed I would rather that it was not, in order that we may give the *Zoo Quest* format a rest and try a new approach, while at the same time preserving the fundamental qualities of the programmes—an escape to far away places.[25]

People of Paradise had originated in an offer from a couple of anthropologists, Elizabeth and James Spillius, studying the people of the Polynesian kingdom of Tonga for the World Health Organisation, to film the Tongan kava ceremony. The anthropologists insisted this ceremony had never been filmed before, thus making 'this a remarkable chance, the sort that only comes once in a generation.'[26] Building on this invitation, Attenborough set up to 'produce an accurate record of traditional life in the island'.[27]

---

[24] *Zoo Quest for the Paradise Bird* (BBC 1957) and *Zoo Quest to Paraguay* (1959).

[25] David Attenborough to Head of Talks, Televisions, 'Expedition to Tonga and the South Seas', 1 May 1959, BBCWAC T6/396/1.

[26] James Spillius to David Attenborough, personal letter, 17 April 1959, BBCWAC T6/396/1.

[27] David Attenborough to James Spillius, personal letter, 5 May 1959, BBCWAC T6/396/1.

This turn to anthropology reflects a change in Attenborough's own interests. Two years later, he signed a new contract with the BBC, under which he worked part-time for the Corporation, devoting the rest of his time to studying for a PhD in anthropology at the London School of Economics (LSE).[28]

Yet, to audiences, Attenborough's presence on the screen did not fade. Quite the contrary, it got stronger. Despite the BBC's efforts to woo new audiences, viewers continued to favour ITV, whose ratings consistently surpassed the BBC's. In this context, Attenborough's programmes were a key part of the BBC's attempt to stem the decline in audiences. And so, although Attenborough wanted to give the *Zoo Quest* format 'a rest', BBC schedulers were keen on maintaining his association with wildlife television present in the public eye, and to keep capitalising on his success with television audiences as well as on the popularity of the *Zoo Quest* series. In 1959–1961, several film compilations and repeats of the expeditions to Sierra Leone, Guiana and Indonesia were broadcast. Attenborough also appeared as a guest in a number of radio programmes to share the sound recordings amassed during his expeditions. For the last *Zoo Quest*, to Madagascar (1961), Attenborough returned to natural history subjects. A ten-episode *Zoo Quest* spin-off was broadcast in the last quarter of 1963. In each instalment of *Attenborough and Animals*, the wildlife presenter introduced extracts from films taken during his various expeditions and displayed live specimens of the animals seen in the films. Audience research for this series paid special attention to the response of children and their perception of Attenborough as a presenter. Although one 'viewer' reported that 'the animals shown … were "rather too rare and exotic" to impress her young children (5–7 age group) who seemed to prefer "the more common animals they had actually seen at the little local zoo"', most viewers were very positive about the programme's content and the presenter. Another viewer reported that their 'three youngsters (boys and girls, 5–11) enjoyed this immensely. Yes, they liked David Attenborough—he is perfect for this sort of programme'.[29]

---

[28] *The Times*, 'News in Brief', 4 August 1962, p. 4.

[29] 'Attenborough and Animals. The first programme in a new series'. Audience research report, 4 November 1963, BBCWAC R9/7/65.

Survival Anglia and the Case for Wildlife Television as Entertainment

The opening of television broadcasting to competition in 1955 had prompted BBC wildlife television to evolve, offering new content and new formats in the hope of attracting new audiences. These, notably, included programmes about travels to exotic places and David Attenborough positioned himself as a major actor in the field. Within the Corporation itself, because of the material means involved in producing programmes based on content filmed overseas, programmes like *Zoo Quest* and *Travellers' Tales* were instrumental in the development of professional practices based on film that could rival, in terms of the scale of the undertakings, those based on outside broadcasting coming from the cultural professional space of radio. At the same time, this film-based approach entailed a much less cumbersome infrastructure. In 1959, Attenborough referred to 'a 16 mm Arriflex with synch sound L2 recorder, together with a 16 mm Bolex in an underwater housing' as 'the standard *Zoo Quest* equipment'.[30] Part of the institutional success of Attenborough's method of wildlife television making rested in its technical simplicity. But establishing a standard which did not rely on the kind of sophisticated and expensive apparatus which the BBC had most chances of controlling—such as the outside broadcast infrastructure—made it easier for others to exploit this concept.

In 1961, a new wildlife series began on British television: *Survival*. Sharply contrasting with what British audiences had been given to see so far, it quickly built a substantive following. It did not originate from the NHU but from a new player in the field: East Anglia Television, one of ITV's regional subsidiaries. In the early 1960s, *Survival* became a driving force for the definition of the style and content of wildlife television in Britain. Its reliance on film and the codes of entertainment cinema, and its rapid gain of financial power through the sale of programmes to American networks also placed *Survival* in a position to exert a selective pressure that favoured some types of wildlife film producers at the expense of others, simultaneously fashioning viewers' tastes and expectations when it came

---

[30]Attenborough to Head of Talks, Television, 'Expedition to Tonga and the South Seas', 1 May 1959, BBCWAC T6/396/1.

to television representations of wildlife. Specifically, *Survival* offered an outlet to a type of wildlife film-making rooted in the imperial culture of big game hunting. The quartet of men who set up East Anglia Television in Norwich in 1958 was a singular mix of competences and interests, which contributed to taking wildlife television further away from the televised lecture format toward film and entertainment. Led by the Marquess of Townshend, deputy lieutenant for the county of Norfolk and a prominent local landowner, the gang of four—as it came to be known—also included Lawrence Scott, editor of *The Manchester Guardian*, film producer John Wolff, whose company, Romulus and Remus, had produced *The African Queen* (1951), and Aubrey Buxton, another Norfolk squire. Mostly known as a skilled birdwatcher, Buxton later became the company's chairman. Other notable associates included the owner of Wyndham Theatres, Donald Albery, and two Cambridge academics, archaeologist Glyn Daniels and anthropologist Aubrey Richards. Each brought to the new channel their personal interests and approach to television.

Anglia Television built its reputation nationally through the production of farming programmes, drama, and series on archaeology, folklore and natural history (Sendall 1983: 11–12). The new channel quickly won audiences locally, with such programmes as *Farming Diary* and *About Anglia*, whose presenter, Dick Joice, was a tenant farmer on the Townshend estate, and nationally, with its drama and television plays. Owing to the channel's connections with the entertainment film world through John Woolf, and with West End theatres through Donald Albery, drama and plays became Anglia's forte. Within a year, several of Anglia TV's plays had placed in the national top ten ratings, regularly getting audience shares of more than 65% nationally, which no programme of any other ITV company had achieved. Through these productions, Anglia made its name as a source of good-quality entertainment. Broadening the audience for their output by broadcasting nationwide was a means for Anglia TV, a commercial company, to increase their profits.

Alongside drama, Anglia also developed a specialism in the production of natural history content. The company's first venture in that area was *Countryman*. Building on his local reputation as a naturalist specialised in birdwatching, Aubrey Buxton fronted the programme, which consisted of films and interviews with local naturalists, with a focus on Norfolk. A

regional equivalent of Peter Scott's *Look*, because of its focus on an amateur naturalist's view of nature, *Countryman* departed nonetheless from
the studio-based format favoured by *Look*. It was mostly film based and
featured Buxton in the field, filmed on location to link filmed sequences
of wildlife, most of which were shot by Ted Eales, the warden of Blakeney
Point (Willock 1978). By the end of 1960, *Countryman* had built a strong
following. Keen to capitalise on this success, Buxton took a programme
prepared for this series and offered it to Associated-Rediffusion as a pilot
for a natural history series to be broadcast nationally on the ITV network.
By then, Granada TV, which thus far had been the main purveyor of natural history content for ITV through their unit installed at the London
Zoo (Chapter 3), had significantly reduced their output, leaving a void
in the ITV schedule. John MacMillan, the controller of programmes at
Associated-Rediffusion, did not like the programme Buxton had brought,
which was about the coypu, a South American species of rodent, invading
the Norfolk countryside. He suggested that Anglia should instead produce
a programme about wildlife in central London.

The London Scene, to be scripted and produced by Colin Willock, a self-
confessed amateur naturalist and author of popular books on the topic,
and shot by film director Bill Morton, Sidney Bernstein's son-in-law, was
the defining first episode of a new wildlife series: *Survival*. Morton, an
accomplished film director, had no interest in natural history whatsoever.
His masterstroke was to commission John Dankworth, a composer well
known in jazz circles, to write a score for the film. Such use of music
squarely positioned the series in the realm of popular entertainment. As
Willock later claimed, this 'appalled the purists but told the general viewing
public that, as far as this show was concerned, natural history had shaken
off the greenery-gallery dust and was in there competing for the ratings
with the best of them' (Willock 1978: 14). To critics, the use of music when
portraying wildlife on television was 'childish'. It was also diminishing the
value of the vicarious experience of being in nature that watching the
programme was supposed to offer. Adding music to the soundtrack was
to 'smudge the veracity, which is the strong point of these programmes'.[31]
In other words, music in wildlife programmes weakened their potential

---

[31] Reginald Pound, 'Critic on The Hearth', *The Listener*, 6 October 1955, p. 568.

for education and edification. At stake in this debate was the essence of natural history and British society's relationship with nature as a source of entertainment or edification.

The innovation which Willock and Buxton claimed to have been theirs with *Survival,* but which to some extent had already being explored in Attenborough's earlier *Zoo Quest*—with the same purpose in mind—was to abandon the notion that entertainment and edification were mutually exclusive and instead use entertainment to expend wildlife television's reach, and therefore its potential for edification. As Colin Willock noted: '*Survival* had deliberately set out to capture the widest possible audience, leaving the specialized wood-notes-wild viewers to the BBC. Aubrey Buxton had declared from the outset that we intended to present wildlife as entertainment' (Willock 1978: 27). Entertainment cinema served as a reference point for the producers of *Survival* (Colin Willock, Stanley Joseph and Aubrey Buxton) both in terms of form and content. They aimed for the production 'of complete documentaries with the standard of editing [found] in a feature film' (Willock 1978: 45). As a consequence, the question of getting film footage became a central one. After two programmes, Willock and Buxton reached the conclusion that, for their enterprise to be sustainable, they needed to assemble a pool of professional wildlife cameramen and help them develop their skills to suit the series' requirements. Many of the most famous wildlife cameramen in the 1960s–1970s worked for the programme and contributed to establishing this standard. Some of them, such as Alan Root, having often had their first independent assignment with *Survival,* subsequently went to work for the BBC. These professional wildlife cameramen, the first of their kind, contributed to establishing in wildlife television a strongly narrative style rooted in both natural historical knowledge and film-making technical expertise. They laid the foundations of a professional culture of wildlife film-making in Great Britain, based on the idea that making a living out of many months spent in the wild filming animals in their natural habitat was indeed possible.

Conclusion

Survival's film-only format sharply contrasted with the NHU's brand of wildlife programming which Peter Scott's *Look* typified, rooted in a sound-broadcasting culture, studio-based and privileging the talk-show format. And although as we saw, when *Survival* appeared, the NHU had already begun moving away from this format under, notably, the influence of the *Travellers' Tales* series (itself an offshoot of the *Zoo Quest* series), the success of the Anglia TV show further encouraged the NHU to change its own style of wildlife programme making. One cause was the growing realisation that Peter Scott's prominence in the NHU's output was an obstacle to attracting viewers beyond the core middle-class audiences he had been very efficient at recruiting in the early days of wildlife television. But these were now over, and a new approach was needed. On the other hand, initiatives in the field of wildlife television, such as the launch of *Survival*, provided the BBC NHU with a convenient straw man against which they could publicly demonstrate their ability to produce educational and edifying content in contradistinction to the 'Pop. Nat. Hist.' which Anglia's productions supposedly epitomised.[32]

References

Attenborough, D. (1955). Expedition to Sierra Leone. *Zoo Life, 10*(1), 11–20.
Castle, K. (2001). Imperial Legacies, new frontiers: Childrens popular literature and the demise of empire. In S. Ward (Ed.), *British culture and the end of empire* (pp. 145–162). Manchester: Manchester University Press.
Durrell, G. (1958). *The Bafut Beagles.* Harmondsworth: Penguin.
Durrell, J. (1967). *Beasts in my bed.* London: Collins.
Gouyon, J. B. (2011). From Kearton to Attenborough: Fashioning the telenaturalist's identity. *History of Science, 49*(1), 25–60.

---

[32]The phrase 'Pop. Nat. Hist.' was coined by Jeffery Boswall, to derisively qualify *Survival* in the early 1960s. See Willock 1978, p. 78.

MacKenzie, J. M. (2001). The persistence of empire in metropolitan culture. In S. Ward (Ed.), *British culture and the end of empire* (pp. 21–36). Manchester: Manchester University Press.

Miall, L. (1994). *Inside the BBC.* London: Weidenfeld & Nicolson.

Parsons, C. (1982). *True to nature.* Cambridge: Patrick Stephens Ltd.

Sandbrook, D. (2005). *Never had it so good: A history of Britain from Suez to the Beatles.* London: Little Brown.

Sendall, B. (1983). *Independent television in Britain: Volume 2 expansion and change, 1958–68.* London and Basingstoke: The Macmillan Press Ltd.

Shapin, S. (2010). *Never pure: Historical studies of science as if it was produced by people with bodies, situated in time, space, culture, and society, and struggling for credibility and authority.* Baltimore: Johns Hopkins University Press.

Teo, H. M. (2001). Wandering in the wake of empire: British travel and tourism in the post-imperial world. In S. Ward (Ed.), *British culture and the end of empire* (pp. 163–179). Manchester: Manchester University Press.

Thomas, J. (2007). *Popular newspapers, the Labour Party and British politics.* London: Routledge.

Waymark, P. (2005). *Television and the cultural revolution: The BBC under Hugh Carleton Greene* (Unpublished PhD dissertation). Open University.

Willock, C. (1978). *The world of survival.* London: André Deutsch.

5

Wildlife Television and Progressivism in 1960s Britain: Rise of the Professional Broadcaster and Downfall of the Amateur Naturalist Film-Maker

In 1961, to self-styled new professional broadcasters, Scott appeared as a remnant of times past, the archetype of an amateurism they desperately tried to escape. In November that year, Eileen Molony, then in charge of producing *Look*, authored a series of scathing internal memos about Peter Scott's unprofessional attitude to broadcasting, whose roots, she said, were plunged in a forgotten age of amateurism and impeded professional broadcasters' work. Scott, she complained, actively 'resisted innovation on past practice'.[1] His multiple engagements, such as refereeing the Olympic games in Melbourne, preparing reports for the IUCN, advising African governments on the conservation of nature, pushed television down his priority list. His chronic lack of availability made it difficult for Molony to fulfil her producer's job efficiently. From the NHU's perspective, Scott was expected to go through at least a complete shot list of the programme before recording the commentary and to arrive, on the day, with a tightly

---

[1] Eileen Molony to HWRP [Desmond Hawkins], 'Peter Scott's contract', 21 November 1961, p. 1. BBCWAC WE8/541/1.

© The Author(s) 2019
J.-B. Gouyon, *BBC Wildlife Documentaries in the Age of Attenborough*,
Palgrave Studies in Science and Popular Culture,
https://doi.org/10.1007/978-3-030-19982-1_5

written script 'à la Attenborough'.[2] Instead, when Scott came to the studio to record commentaries, his lack of preparation was evident. To Molony's dismay, if asked to prepare such notes, Scott replied that 'he did not feel qualified to do so'.

> On certain days he sits in the studio talking without ceasing, improvising ideas, criticising design, etc., whilst one is trying to communicate through the same loudspeaker with studio manager, telecine, and Studio B. He has no idea of putting a precisely prepared question and allowing the guest to answer, so that one can cross cut in a polished way, but dribbles on unpredictably so that a studio situation has to be created in which a two-shot of both speakers is always available.[3]

Molony acknowledged that, beyond his following, Scott brought many qualities to *Look*. But they belonged to the culture of amateur natural history, and Scott desperately lacked any of the character traits expected from a professional broadcaster. His visual memory and his 'feeling for film', for example, were invaluable, as well as his 'generalised natural history background'. But he operated with outdated conceptions of television as a medium and did not seem particularly interested in it. Molony wished that Scott would watch 'a little more television and see how other compères are treating their material as television advances.'[4]

Molony's critical evaluation of Peter Scott's performance is emblematic of the period of reflexivity which began at the NHU in the early 1960s. As *Survival*, the wildlife programme from Anglia television, thrived, audiences for the BBC NHU's output plummeted. This prompted anxious self-examination on the part of the staff in Bristol, leading to reconsidering which type of programmes to produce as well as how to do it. Molony's assessment of the situation with Peter Scott further reveals the issues and solutions identified at the NHU. Broadcasters in Bristol understood the main source of their difficulties to lie in their overreliance on the culture

---

[2]Eileen Molony to HWRP [Desmond Hawkins], 'Peter Scott's contract', 21 November 1961, p. 2. BBCWAC WE8/541/1.
[3]Eileen Molony to HWRP [Desmond Hawkins], 'Peter Scott's contract', 21 November 1961, p. 1. BBCWAC WE8/541/1.
[4]Eileen Molony to HWRP [Desmond Hawkins], 'Peter Scott's contract', 21 November 1961, p. 3. BBCWAC WE8/541/1.

of amateur natural history, of which Peter Scott was—rightly—perceived to be a standard-bearer, and on the spoken-word culture inherited from radio, where his encyclopaedic knowledge was well suited to improvised, live commentary in a studio setting. By contrast, Molony and her colleagues engaged in developing a professional culture of wildlife television production, aiming for a closer collaboration with scientists and increasingly relying on pre-recorded films, as opposed to live studio discussions.[5]

The Unknown Forest: The NHU Starts Experimenting with the Codes of Entertainment Cinema

In 1961, as ITV was consolidating its wildlife offer, becoming, with *Survival*, more visible in the television landscape, it was also gaining greater audience shares, especially among 'the great mass of viewers … available in the industrial areas of the Midlands and the North' (Parsons 1982: 263). In other words, ITV was more successful than the BBC with viewers from the working classes. This spread of the television medium through the social body led the Bristol wildlife broadcasters to conceive of *Look* and its star presenter, Peter Scott, as a liability with regard not only to his presenting style and his understanding of television but also to socially diversifying their audiences. Sailing with King George VI, sketching portraits of Princesses Elizabeth and Margaret in 1947, he was socially acquainted with the British royal family. Once invited to a shooting weekend at the royal residence of Sandringham, he showed his hosts the film of his expedition to Canada (Scott 1966). Scott, like many of his guests in the early *Look* programmes, belonged to the upper classes. A war hero, he was the quintessential English gentleman, who could lead an unconstrained life of rational leisure. In an evolving cultural landscape where television audiences were growing increasingly socially diverse, Christopher Parsons and others in the NHU were quick to figure out that the prominence of Peter

---

[5]This transition from a studio-based format to a film-based one was also central to early discussions about the BBC science programme *Horizon*, between 1964 and 1966 (Boon 2015).

Scott and *Look* in their output was a potential weakness. Scott's audience was very homogeneous, which limited its potential for growth and evolution. To address this issue, the NHU set about developing wildlife programmes with features that clearly distinguished them from *Look*. New presenters with a broader social appeal were brought in at the same time as the NHU experimented with the production of features films that imported codes of entertainment cinema to wildlife television.

A key participant in this endeavour was Johnny Morris (1916–1999), who used different voices to express animals' thoughts and emotions. Parsons and his colleagues believed that this very distinctive style of presentation resonated with popular audiences. Born in 1916, in Newport, Morris grew up in a family of amateur musicians and often accompanied his father to perform in pubs. In 1945, Morris met with Desmond Hawkins, then a Radio Features producer at BBC West. Taken by Morris's gifts 'as a mimic, a master of language, and vocal inventor of his own sound effects', Hawkins invited him on the radio to feature in light entertainment programmes.[6] In 1957, Morris went on television to narrate the English version of *Tufty*, a film by Swedish director Bertil Danielsson (1914–1982), whose main character was a tufted duck. Initially broadcast as a Christmas special, the film was repeated several times afterwards, owing to popular demand. Following this inaugural success, Morris and his producer, Tony Soper, embarked on a long-term collaboration, the highlight of which was to be *Animal Magic* (1962–1984). Audience research found that Morris's empathetic style of presentation, described as a blend of humour, compassion and emotion, appealed to a wide range of age groups. Of special significance to NHU executives, Morris had enthused and encouraged viewers who usually did not watch *Look* to engage with a natural history film on television.

When Christopher Parsons began looking for a voice to narrate another of his projects, *The Unknown Forest*, the first long television feature to come out of the NHU, he turned, initially, to Morris to write and deliver the commentary. Parsons's declared objective was to use Morris's appeal to attract viewers with no interest in wildlife and show them what, at the

---

[6]Tim Bullamore and Desmond Hawkins, 'Obituary: Johnny Morris', *The Independent*, 8 May 1999. Available online at https://www.independent.co.uk/arts-entertainment/obituary-johnny-morris-1092133.html. Last accessed 18 March 2019.

time, was considered the essence of the NHU's specialist output: films of British wild animals behaving as if the cameraman were not there.

> It was really kicking against the traces, I just felt that if we were to compete with ITV we ought to be doing something that was a bit different from *Look* and trying to sort of grab a family audience and making something that was a bit more, I don't know, would have some kind of resonance with you know a family in Coronation Street or something like that.[7]

The Unknown Forest, broadcast on 19 January 1961, was meant to shake up the conventions of wildlife television. In line with Parsons's vow that it should 'entertain as many people as possible' and represent 'a new kind of wildlife film' (Parsons 1982: 108), it features many characteristics of entertainment cinema, not found to such a degree in previous BBC wildlife television programmes. Right from the opening credits, the storytelling intent is made clear: Morris is presented as the storyteller. The action then opens on a sequence featuring fictional picnickers driving through the New Forest in southern England and setting up for a rest under the gaze of an owl. Besides Morris's engaging, unashamedly anthropomorphic commentary, the soundtrack departed from the then compulsory focus on natural sounds. Based on a score in 'the pseudo-pastoral English style',[8] it had been especially composed by trombonist Sidney Sager, a friend of Morris and, at the time, the conductor of the BBC West of England Light Orchestra. The film overtly blurs the distinction between facts and fiction, offering itself to viewers as a truthful recreation of wild nature to be enjoyed not only for the purpose of acquiring knowledge but also for the sensual entertainment it provides. In a 2001 oral history interview, Parsons commented: 'It actually shook things up a bit. And made people realise that there were other approaches.'[9] Yet, despite an enthusiastic reception from audiences, Parsons and others at the NHU were aware that if they ventured too far in the entertainment direction, they risked losing 'the respect and co-operation [the Unit] enjoyed with many noted

---

[7]Wildscreen, 2001, 'Christopher Parsons oral history interview', 26 July 2001.
[8]As described by *The Times* television critic, 'Rich nature near the beaten tracks', 20 January 1961, p. 4.
[9]Wildscreen, 2001, 'Christopher Parsons oral history interview', 26 July 2001.

naturalists and zoologists' (Parsons 1982: 263), for whom the only valid justification for presenting nature on television was to convey knowledge about it. To pre-empt criticism, the BBC focused its communication on the personality and methods of the amateur naturalist cameraman who had shot the images shown in *The Unknown Forest*, Eric Ashby (1918–2003).

In 1961, Eric Ashby, a smallholder at Ringwood near the New Forest, came to epitomise amateur natural history film-making at its best: a combination of a high degree of expertise in natural history with the technical skills and patience to obtain high-quality footage of animals in the wild. His work had first come to the notice of the NHU through a short article published in *The Countryman* detailing how he had photographed wild deer in the New Forest and also mentioning some 16 mm footage. Roger Perry and Christopher Parsons, from the NHU, had visited Ashby to see his film work and found that he had developed a specialty of obtaining close-ups reputedly impossible to get, such as shots of wild badgers in broad daylight. These were evidence of intense field craft, patience and dedication. Most of Ashby's footage could only have been obtained after protracted sessions of stalking animals, taking care to approach downwind and avoiding producing any noise that could alert them. Essentially, Ashby had been standing perfectly still in the woods for hours on end. Sometimes, an entire evening of work resulted in twenty seconds of film. In the field, Eric Ashby deployed the skills of a hunter, 'pitting *his* [original emphasis] senses against the animals'' (Parsons 1969: 18). His footage fitted well with the rhetoric of progress associated with wildlife film-making: the never-before-seen trope. Previous films of badgers shown in *Look* had been obtained by filming the animals at night, after they had been habituated to artificial lighting. Here was the promise of enjoying an encounter with nature more unmediated than ever, thanks to the work of this new, never-heard-of-before amateur naturalist cameraman.

The cognitive value of Ashby's footage was further reinforced by his modest demeanour, his unwillingness to attract attention and be placed under the spotlights. As the readers of the *Radio Times* were told, Ashby was 'a rather shy smallholder' only interested in spending the spare time farming afforded him 'stalking and filming' wild animals. That Ashby had remained unknown thus far, despite all his skill, experience, patience and determination, was evidence of his modesty, and thereby, his trust-worthiness. As historians of science Steven Shapin and Simon Schaffer (1985) showed, displays of modesty were integral to truth telling in the seventeenth century, when the foundations of modern experimental science were laid. In the early 1960s, as in the 1660s, modesty was part of an effort to erase the naturalist's subjectivity and emphasise the collective effort implicated in the production of objective facts about nature. In addition, the description of Ashby as self-effacing should be taken literally for it points to his ability to produce accurate representations of wildlife as if the camera were not there—Ashby, himself the camera holder, achieving invisibility. His footage was thus endowed with a high degree of index-icality: it could be claimed to have a strong relationship of equivalence with the elements of reality it was supposed to represent. To insist on Ashby's shyness was to locate his film in the repertoire of observational realism, which enjoyed some currency in the early 1960s with the development of cinema vérité. Conversely, this claimed indexicality downplayed the significance of *The Unknown Forest*'s entertainment dimension, which manifested itself in the score, in Johnny Morris's storytelling, and in the sequences featuring fictional characters. The evidence that the film could be trusted as a reliable source of knowledge of the natural world had become embedded within the film itself, as made clear by the publicity around it, revealing how it had been obtained. With the constitutive elements of the film guaranteeing its epistemic reliability, the form it takes becomes almost irrelevant. Notably, it becomes possible to adopt some of the codes of entertainment cinema, previously thought to be alien to this type of television content.

Parsons's plans were to use Ashby's films as material to develop an alternative to *Look*. Ashby's own ambition, however, was to appear in Peter Scott's series, and so it is there that he featured most significantly afterwards. Furthermore, the very slow pace at which Ashby was able to film rendered any attempt at producing a feature based solely on his footage a protracted process. It had taken him four years to obtain the footage used in the forty-five-minute *The Unknown Forest*. In the *Look* episodes featuring his work, several sequences showing Ashby with badgers or foxes playing around him, unsuspecting of the human presence, further reinforced the fashioning of the cameraman as a modest witness. Here was evidence that his body could become transparent, that he could be present on the scene without disturbing it, and thus trusted to produce objective representations of nature. These sequences defined Ashby as an unobtrusive presence in the wild, suggesting that he could become part of the landscape. As Peter Scott explained in *Forest Diary* (1963), they demanded a lot of preparation and staging, and depended on conditions being right. For example, Ashby had to make sure that he stood perfectly still and downwind from the animals so they would not detect his presence. Such disclosures far from undermining the value of the sequences, further reinforced the notion that the film-maker possessed enough intimate knowledge of nature to be able to blend into it, to become nature. In the context of an increasingly aggressive competition, notably from East Anglia television, the presentation of Eric Ashby as a 'modest witness' producing truthful accounts of nature with his camera can be interpreted as an attempt by the NHU to characterise its own wildlife output as epistemically authoritative. However, Ashby can also be seen as the last hurrah of the culture of amateur natural history in relation to wildlife television. Although he would continue to produce footage for the NHU well into the 1980s, the NHU stopped putting him forward and promoting him as a key figure after 1962–1963. The age of the amateur wildlife cameraman was coming to an end; the professional was displacing the amateur as the figure commanding expertise.

What Is a Baby? Importing the Attenborough Style to Bristol

In 1962–1963, a vanguard of wildlife broadcasters was questioning the value of live, studio-based sequences in programme making. Conceiving of themselves as experts, able to skilfully manipulate pre-recorded material in order to knowingly produce desired effects, they had begun developing a professional culture of wildlife television making centred on postproduction. The editing room and the dubbing theatre were superseding the studio as the core locations of programme production. From this vantage point, studio hosts' performance was not the only one that mattered anymore. Another type of performance, the delivery of a commentary as part of the production of a film's soundtrack, was becoming essential to the success of a television programme. This approach rested on a desire for quality control. It facilitated the emergence of technical standards of production around which a professional community of producers could assemble. But because, save for Christopher Parsons, who had started his career in television, the first generation of wildlife programme makers in Bristol all came from radio and had merely transferred the studio-based live format to television, adopting this approach required a major cultural shift to take place at the NHU. Alongside Parsons's venture into producing television feature films such as *The Unknown Forest*, bringing in David Attenborough to participate in programme production helped coax the NHU in this direction. Since 1956 and the introduction of commercial television in Britain, the *Zoo Quest* series had consistently been the most successful BBC wildlife programme with audiences. Broadcasters in Bristol thus perceived Attenborough's programme making style, which notably included the use of 'tightly written scripts' for delivering voice-over commentaries and prioritised pre-recorded material over live sequences, as a recipe for success.

Attenborough's first formal collaboration with the NHU was in 1962 alongside Peter Scott in the programme *Guest at Slimbridge*, set in Scott's home and natural reserve, where he kept his collection of wildfowl. The programme was billed as a summit meeting, bringing together for the first time what the *Radio Times* described as the 'two leading zoologists' of British wildlife television.[10] Less than a year after his visit to Scott's territory—described by Jeffery Boswall, a fine observer of the professional milieu where he evolved and famous for his witticisms, as 'introducing a fox into the water-fowl pens'[11]—Attenborough began working with Eileen Molony. As the producer of *What Is a Baby?*, a programme comparing humans with other animals, she initially asked Attenborough to build on his knowledge of anthropological films and suggest source material representing 'children up to about 2 years'[12] in different cultures from the extensive collection amassed through the *Travellers' Tales* series. She

---

[10] Jeffery Boswall, 1962, 'Guest at Slimbridge', *Radio Times*, 3 May 1962, p. 34.
[11] Ibid.
[12] Eileen Molony to David Attenborough, personal letter, 5 June 1963, BBCWAC WE8/83/1.

also invited him to write and record a commentary for the whole thirty-minute programme. As well as wildlife and anthropological footage, the programme was to include an interview by Attenborough of an 'affable and extroverted' paediatrician.[13] Attenborough got closely involved in the production process from the outset, providing advice on the choice of stock footage, on the direction the programme should take, and on practical aspects of the production. Several letters exchanged in the process portray an emerging relationship between Attenborough and the NHU, notably through the mutual construction of agreement over matters related to the professional practice of producing television programmes.

One lesson Attenborough had taken from a decade of learning the ropes was that only pre-recorded material could afford the degree of precision required for successful television programme making. Thus, writing to Sheila Fullom, Eileen Molony's secretary at the NHU, Attenborough extolled the merits of pre-recorded sequences over live studio ones:

> I share your view of film versus studio. To begin with I got a tremendous kick out of the excitement of putting out programmes live. But it wore off after a bit & really, except for challenging interviews with lots of 'immediacy', I'm for film or some other sort of controlled recording process every time. It is so maddening to miss an effect because of some small mechanical hitch, as so often happens live. That's why I enjoyed making the Japan film so much. The raw material was all unstaged actuality, but we were able to manipulate it with precision later.[14]

This approach establishes post-production, notably editing, as the key moment of programme making, defined as the assemblage of a tightly controlled visual performance, intended to produce specific effects on viewers.

Key to this performance was the presenter's commentary, which, in keeping with the degree of control pre-recorded sequences offered, was to result from thorough preparation work, and as such, could not be improvised. When the time came to record the commentary for *What Is a Baby?*, Attenborough sent back his annotated version of the script. In the

---

[13]Eileen Molony to David Attenborough, 5 July 1963, BBCWAC WE8/83/1.
[14]Attenborough to Fullom, personal letter, 1 August 1963, BBCWAC WE8/83/1.

cover letter, he emphasised the performative aspect of his contribution, and insisted on the use of a teleprompter to guarantee the quality of the performance:

> As for the vexed problem of Autocue: I must be candid and admit that it would be a great relief to have most of the expository bits on Autocue ... This is not entirely laziness. The programme, as it is now scripted, is, after all, the result of considerable thought, and the alternative seem to me to be either to learn it word for word or else to half ad lib it, in which case one risks not getting the emphases exactly right and the whole thing losing its tightness and precision. Furthermore, if we use Autocue, I suspect that Eileen will have much more confidence in her somewhat erratic speaker, will know exactly where he has got to, will be certain of where to cue telecine, and will be able to change even words and inflections—which is I fear with me, something of a toss up [sic.] otherwise.

> I realise, of course, that I may be speaking to more than one camera, but I believe Autocue can have two machines working synchronously on two cameras, so that there is no difficulty in changing from one to the other in mid-sentence if necessary.[15]

The letter can be read as a reply to Molony's earlier complaints about Scott's way of approaching his role as a studio host. There Attenborough presents himself as a professional broadcaster, keenly aware of the latest advances in television technology. Autocue, a British company specialising in producing teleprompting devices, had only licensed its first on-camera teleprompter in 1962. While appearing to simply describe the division of labour between the producer and the presenter, Attenborough in fact defines it as a co-produced performance, whose value and 'precision' rest on the research and preparation work he put into producing the script for the programme. These two quotes sketch the main lines of the professional culture of television broadcasting emerging in the early 1960s in Bristol, which relied on two kinds of expertise: technical, to handle broadcasting equipment and adapt to the constraints it imposes; thespian, to prepare for and deliver seamless visual and oratory performances.

---

[15]David Attenborough to Sheila Fullom, personal letter, 17 July 1963, p. 1. BBCWAC WE8/83/1.

The Alder Woodwasp and Its Insect Enemies: Taking Nature Apart and Putting It Back Together Again On-Screen

The emergence of this new professional culture of broadcasting, placing the emphasis on post-production, created the need for a steady incoming flow of raw film material of reliable technical quality. A key, if only momentary, partner in this endeavour was the Council for Nature, an advocacy group established in 1958 'to revive the natural history movement in Britain', and 'create an active public opinion' in favour of natural history across the country.[16] In collaboration with the Council for Nature, the NHU thus engaged in several initiatives destined to develop a pool of wildlife camer-amen working to the required standard. These were, for the NHU, part of a concerted effort to model an ecology of wildlife film-making in Britain around shared standards of practice and where the Bristol NHU, as a cen-tre of excellence, could stand in a position of dominance. In 1960, the first of these joint initiatives was to launch a co-sponsored natural history film-making competition. In addition to a £500 prize, the winner would get their film broadcast on the BBC. Entries were limited to film-makers who had original films with no more than ten minutes of their footage previously shown on television in the UK, which de facto excluded all pre-vious *Look* contributors, most of them members of Peter Scott's network. The competition rules instructed participants on the speed (24 fps) and gauge (16 or 35 mm) at which to film. In the early 1960s, amateur nat-ural history film-making in Britain was characteristically diverse in terms of practices and standards. For instance, no particular filming speed pre-vailed, many amateurs shooting at speeds other than twenty-four frames per second, especially when they shot silent films. Likewise, 8 mm film, because it was inexpensive, was popular. Besides providing the NHU with original film material to include in their programmes, the competition helped spread technical standards of practice in the community of ama-teur wildlife cameramen.

---

[16]E. M. Nicholson to Desmond Hawkins, personal letter, 7 July 1958, The National Archives, FT 3/541.

The winning entry in the 1960 competition, Gerald Thompson (1917–2002) and Eric Skinner's *The Alder Woodwasp and Its Insect Enemies*, became a benchmark in film-making practice and sophistication. It also contributed to ushering in yet a new approach to wildlife television production whereby film-makers took nature apart only to reassemble it on-screen to fit a predetermined storyline. The film depicts how the alder woodwasp deposits its eggs between the bark and the wood of the alder tree, and shows a succession of four parasitic insects laying their eggs in or next to the developing pupae. Received with great excitement at Bristol, the film was praised as 'an excellent film with outstanding CUs [close-ups] of a difficult subject. The whole story is extremely well covered and imaginative in treatment, apart from being technically good.'[17] With this film, hailed as breaking new technical grounds for its use of macro-cinematography techniques, Thompson, a lecturer at the Forestry Institute at Oxford University, and Skinner, his assistant, not only reiterated Heinz Sielmann's tour de force with the woodpeckers five years earlier but took it a step further, for most of the filming took place in a controlled environment.

To shoot *The Alder Woodwasp and Its Insect Enemies*, Thompson and Skinner had to overcome a number of practical problems. These included finding a way of dealing with the great heat generated by the powerful lighting needed for macro-cinematography. Heating the air to around 70 °C, the light coming from the 500-watt bulb used to illuminate the film set is lethal to insects. To cool it down, Thompson and Skinner combined a three-litre flask filled with distilled water, which acted as a primitive condenser, with two filters cut in a special hardened glass which could absorb almost 85% of the infrared light. Another potential obstacle, and a crucial one, was that all the behaviour to be filmed was happening on tree trunks in the space between the wood and the bark, hidden from unaided human sight. Apart from establishing frames taken in the forest, the rest of the film was shot in a studio, with wood, wasps and parasites brought in and set up so that the hidden processes of the wasp laying its eggs and the larvae being parasitized could be captured on film. To film the events

[17]NHU, 'Summary of films viewed in the Natural History Unit November 1960–March 1961', p. 15. BBCWAC WE17/2/1.

taking place between the wood and the bark, Thompson had to repeat a very delicate manipulation thirty-two times before achieving success. After a wasp had landed on a piece of alder stem and begun drilling a hole in the bark with its ovipositor, Thompson used a scalpel to cut through the bark around the insect. If he had not disturbed the wasp, he could then lift the piece of bark, without altering its natural curvature. If Thompson failed, the ovipositor was pinched, and the wasp withdrew. If all went well, he could film the normally invisible process of the eggs emerging from the ovipositor.

To represent nature faithfully, Thompson and Skinner took it apart, only reassembling it on-screen through the editing process. But the version they obtained, although accurate, was unavailable in the wild. Whereas amateur natural history cameramen like Eric Ashby, for instance, strove to produce representations of nature that negated intervention, Thompson and Skinner produced representations of nature which avowedly could not have been obtained without intervention and, in fact, magnified it. Their main conundrum had been to identify the degree of intervention that would leave intact the natural phenomena they wanted to film and to create the artificial filming conditions under which the insect could act naturally. This interventionist approach to filming fitted well with the idea, which the NHU was increasingly keen to put forward, that film-making could be a participation in the production of knowledge about the natural world. The billing for the programme in the *Radio Times* emphasised this notion when claiming that two of the insects' egg laying behaviours depicted in the film were 'new to science'.[18]

Unarmed Hunters: Bristol Wildlife Broadcasters Come Out as Professionals

As part of their effort to create a critical mass of wildlife cameramen working to the NHU's required standard, the NHU again collaborated with the Council for Nature, this time to deliver a training programme on wildlife film-making. Consisting of a series of short courses on film-making

---

[18] *Radio Times*, 6 April 1961, p. 57.

and sound recording techniques,[19] the intended audiences were members of local natural history societies. The NHU's contribution took the form of teaching material, a set of notes on technical aspects of film-making and a film presenting how professionals worked at Bristol. This teaching material did not merely impart technical knowledge on how best to film wildlife; it granted pride of place to film editing as the pivotal moment in the production of wildlife television programmes. The 'Notes on Filming' distributed to participants in the course encouraged the production of streamlined raw material that would make any subsequent work in the cutting room easier. Besides the request to shoot at twenty-four frames per second, trainees were asked to think in terms of stories, composed of successive sequences linked together pictorially:

> Give the subject as much variation as possible—wide angles for geographical and setting purposes, and mid-shot for coming in close to a subject. Because of the size of the screen, close-ups are required for television, but not a quantity unrelieved by anything else. Please give a change of angle to the shots on the same subject, together with occasional changes of distance from the subject.[20]

This document was intended to foster the NHU's desired mindset in would-be wildlife cameramen. To ensure consistency of style, the document instructed that a tripod should be used, and that the question of 'scale-relative size' should be kept in mind in the form of the inclusion of such artefacts as trucks or cars in frames showing 'towering mountains'. When panning, the camera should be kept 'running for, say, eight feet [of film]' before and after panning is finished as '[t]his gives the editor a good deal of leeway'. Editing is placed here at the forefront; aspiring wildlife cameramen had to imagine their work primarily as producing material that could be edited. As much as stipulating the form filming should take, these 'Notes on Filming' equally imposed constraints on the desired content of the footage. The NHU producers were primarily looking for life

---

[19]Bruce Campbell, 'The Natural History Unit of the British Broadcasting Corporation', March 1962, BBCWAC WE17/2/1.
[20]Natural History Unit, 'Notes on filming', 9 February 1962, BBCWAC WE17/2/1.

stories focusing on one individual organism from birth to death, comparative studies of patterns of behaviour across different groups of animals, or habitat studies 'revealing as much as possible the inter-dependence of the different forms of wild life' found for example at the edge of the jungle.

This whole set of instructions suggests a high degree of formalisation of the practices involved in filming wildlife, from purely technical aspects to the choice of subjects. It is symptomatic of a will to rationalise the production of television programmes and establish a division of labour revolving around programme producers working in tandem with film editors. In this new ecology of wildlife television, cameramen lose some of their agency as they become defined as suppliers of the raw material that will enable programme producers to create truthful representations of nature. This pre-eminence of the producer-editor tandem in the production of wildlife television programmes was, likewise, at the heart of the NHU's main contribution to the Council for Nature's training course in wildlife film-making: a teaching film. *Unarmed Hunters*, a making-of documentary, of a sort, was shot in late 1962 and early 1963 and took trainees through a virtual visit of the NHU, explaining how it functioned. It was produced and directed by Christopher Parsons, the main advocate at the NHU of the use of film as opposed to live studio programmes in wildlife television. It is therefore no surprise that *Unarmed Hunters*, in its depiction of twenty-four hours in the life of the NHU, should place the emphasis on post-production.

Shot in black and white, in fly-on-the-wall style, *Unarmed Hunters* is indebted to the direct cinema movement. It was not simply about disclosing the different aspects of wildlife television production; these disclosures served to normalise a style of filming. Direct cinema shares with wildlife film-making a belief in observational realism. Both claim to collapse the difference between image and reality, and to offer unaltered representations of the world. In both cases, non-intervention on the part of the film-maker is highly valued, and the camera is conceived of as a means of recording events that would have occurred whether the camera was present or not, thus supposedly providing access to an objective, unmediated reality. Just as in stories told with no clearly identified narrator, 'events seem to tell themselves'[21] in documentaries shot in the direct-cinema style. This style of filming is therefore a powerful tool to naturalise and thus normalise the actions depicted on-screen. *Unarmed Hunters* is a case of wildlife film-makers applying to themselves their own naturalistic approach to film-making.

The film is a celebration of machines and offers a very technical definition of wildlife television production. *Unarmed Hunters* is mostly composed of two kinds of sequence: people at work and the equipment they use. The latter offer close-up shots of machines and various pieces of equipment. As a film meant to instruct potential future contributors to the NHU, *Unarmed Hunters* introduces viewers to the successive stages of programme making. From initial search for existing stock footage in the NHU film library to dubbing to finalise the sound-track with additional pre-recorded sounds, through to the rehearsal and shooting of link sequences, the film emphasises the manufactured nature of wildlife television programmes. Throughout, the camera lingers over Steenbecks, stacks of reels, telecine machines, mixing consoles, quarter-inch tape decks, and hands pushing buttons and turning knobs. The abundance of machinery, often looking complex, liken the production of wildlife television programmes to an elaborate industrial process, which only technical experts can adequately perform. One film-maker fitting this model was Gerald Thompson, whose work is presented in *Unarmed Hunters*. Parsons's aim when filming Thompson, was 'to show the mass of precision equipment necessary to get the results that you do, and to give some indication of the time spent'.[22] The sequence shows Thompson and Skinner at work and condenses some of the action involved in obtaining one of the close-up shots of insects for which Thompson was building a reputation. By contrast, Eric Ashby, filmed from afar as he made his way through undergrowth, carrying his 16 mm camera in a wooden box, appears alien to the machine-saturated atmosphere of *Unarmed Hunters*. Dressed up for field work in tweeds and wellies, his demeanour further distances him from

[21] E. Benveniste quoted in H. White, 1980, 'The value of narrativity in the representation of reality', *Critical Inquiry*, 7(1), 5–27, 7.

[22] C. Parsons, personal letter to Gerald Thompson, 28 February 1963, BBCWAC WE21/68/1.

NHU workers, clad in suits and looking more like civil servants than gentlemen farmers.

In *Unarmed Hunters*, amateur naturalist cameramen are pushed to the margin of this renewed ecology of wildlife television production. By placing such emphasis on broadcasters' professionalism, the Bristol NHU was falling in line with the rest of the Corporation. Interviewing BBC workers in 1963, sociologist Tom Burns found out that in the institutional culture then prevailing there, such words as 'professional' or 'professionalism' had become central, as ways of claiming and supporting claims to excellence and importance (Burns 1977: 123–124). And just as was the case at the NHU, this prevalence of the notion of professionalism went hand-in-hand with the spread of technical specialisation within the BBC at large. As Burns remarks, such self-styling as professionals enabled BBC workers to disqualify outsiders as unable to perform the job correctly. These outsiders lacked expertise; they were amateurs. The professionals working at the NHU emerge from *Unarmed Hunters* as experts in the production of reliable and accurate representations of nature.

A key sequence in the film offers a behind=the-scene view of the final stage of dubbing for a *World Zoos* series episode about Helsinki's zoo. The sequence features Nicholas Crocker, who was then the editor of the NHU, his personal assistant, Pamela Everett, and film editor George Inger. Dubbing is explained in the voice-over commentary: 'The various soundtracks of music, sound, speech and effects, which have been carefully synchronised by the editor are now mixed together in the right proportions'.[23] The sequence includes a discussion about the level of traffic noises in the opening of the programme. Crocker remarks that since the zoo is located in the centre of Helsinki, they are 'a bit thin' and stronger sound effects should be added to the soundtrack. This sequence dispels the notion that wildlife television programmes offer an unmediated encounter with what they represent. On the contrary it insists that for programmes to be accurate representations of the world, a necessary process of re-assemblage has to take place during post-production. The editing room and the dubbing studio become places were nature is produced. The reality of nature stands here as the outcome of the film-making process, which becomes a means

---

[23] *Unarmed Hunters* (BBC-NHU, 1963).

of knowing nature. *Unarmed Hunters* defines the production of filmic representations of nature as the raison d'être of natural historical enquiry.

Besides being shown on television in November 1964, *Unarmed Hunters* served to present the NHU in a variety of contexts, from natural history festivals to meetings of natural history societies. The claims about the wildlife film-making it encapsulated thus reached a larger audience than the cohorts of aspiring wildlife cameramen taking the Council for Nature's film-making courses. Internal correspondence referred to the film as a 'BBC demonstration film'.[24] In the late 1930s, and again in the early 1950s, the phrase had been used to designate films intended to promote the television service to help sell television sets. These demonstration films were broadcast before the start of the actual schedule so that television-set sellers had something to show to prospective buyers. They had a typically low budget as they consisted of compilations of extracts from past programmes interspersed with test cards and specially shot announcements. The NHU broadcasters conceived of *Unarmed Hunters* as a promotional tool, but it contained very little existing programme material, instead mobilising significantly more costly resources than a typical demonstration film. To commit film rolls, filming and editing equipment, as well as the people to operate them, a speaker for the commentary and a producer, all travelling to different filming locations such as Thompson's studio in Oxford, suggest that the project was considered an important element in the NHU's communication about itself. After the film had been produced, the NHU ensured that they retained complete control over circulation by taking and paying 'for full N/T [non-theatrical] rights (and T.V. rights)'.[25] In 1964, *Unarmed Hunters* was unique in this respect, and the only film in the NHU's library which the NHU could circulate as they pleased with no need to get permission from Television Enterprise, the BBC department routinely charged with commercially exploiting the Corporation's programmes. Owing to such freedom, *Unarmed Hunters* was publicly shown numerous times in the years that followed its production, notably in local natural history societies. For instance, on 24 February 1966, the

---

[24]N. Crocker, 'Natural History Unit library', personal letter to A. T. Callum, 7 August 1964, BBCWAC R125/829/1.

[25]N. Crocker, 'Natural History Unit library', personal letter to A. T. Callum, 7 August 1964, BBCWAC R125/829/1.

NHU's sound librarian, John Burton, introduced it at a meeting of the South London Entomological and Natural History Society. To circulate *Unarmed Hunters* in the amateur natural history community was meant to convince practising naturalists of the relevance of film-making for the production of valuable natural historical knowledge, and to uphold the NHU's standing within this community (Gouyon 2016).

Reaching an even larger audience, *Unarmed Hunters* was one of the films shown during the 1963 National Nature Week in the cinema the BBC had installed on this occasion in the Old Hall of the Royal Horticultural Society in Westminster. The 'BBC's special cinema' was part of a larger exhibition sponsored by *The Observer*, the centrepiece of National Nature Week, a week-long festival celebrating nature and the natural history movement across Britain, which a total of 46,160 persons visited. As Bruce Campbell explained in *The Guardian* in May 1963, during National Nature Week, 'the naturalists of Britain are out to affirm their faith and show their work to their countrymen at a time when our wildlife heritage is in the balance'.[26] By participating in this festival, the NHU presented itself as an active member of the British community of natural history, whose films could foster an interest in wildlife. The entire programme of the BBC cinema in the RHS Hall, featuring recent productions of the NHU, emphasised the value of film-making as a mode of engagement with nature and as a process to construct representations from which viewers could derive knowledge of the natural world. These films were meant to enable those with no time to devote to the study of natural history to nonetheless experience the 'enormous pleasure' that 'looking at unspoiled scenery and wildlife' can provide.[27] Alongside *Unarmed Hunters*, the BBC showed a number of other wildlife films, *The Unknown Forest*, the two winners of the second natural history film-making competition, and *The Major*.

The first film originating from the NHU to be shot in colour, *The Major* tells the story of an old oak tree planted in the middle of a village. From villagers using it as a meeting place and a notice board, to tiny gall-wasps and large caterpillars, through to birds and small mammals, the film is a close study of the various ecological communities gravitating around

---

[26]Bruce Campbell, 1963, 'A mirror up to nature', *The Guardian*, 18 May 1963, p. 6.
[27]Peter Scott, 1963, 'National nature week', *Radio Times*, 16 May 1963, p. 6.

the oak. It is also a synthesis of two types of wildlife camerawork. Eric Ashby's traditional amateur naturalist film work, in the wild, stands side by side with specialised filming under controlled conditions. This latter footage came from Leslie Jackman, a school's museum officer at Paignton Aquarium in Devon. Jackman had developed a specialty of filming small life forms in aquariums or vivariums. He was part, like Thompson, of the new contingent of external contributors the NHU was nurturing for their ability to produce footage of professional standard. To juxtapose in *The Major* the two types of footage was to demonstrate the equal value of both to an accurate reconstruction of nature on-screen. It makes the case for the necessary artificiality of the film-making process, placing it at the core of wildlife television.

The framing of *The Major* in the *Radio Times* before its transmission on 20 May 1963, as part of the broadcaster's special schedule for National Nature Week, emphasises the constructed nature of the film. Contrary to most natural history programmes, the idea for *The Major* had come from the producer for the film, Christopher Parsons, rather than from an amateur naturalist film-maker approaching the NHU. The billing in *Radio Times* does not simply summarise the film but insists on the way it was made. It reveals that Parsons and John Burton wrote the script for the film even 'before beginning to look for their tree. Eventually the Forestry Commission found one that satisfied nearly all their requirements'.[28] The text then proceeds to detail for readers the film-making process, illuminating how 'producer Parsons and cameraman William Morris visited the tree to record various activities throughout twelve months', all while Eric Ashby and Leslie Jackman were filming commissioned sequences to illustrate other portions of the script. To emphasise the production process implicitly normalised the notion that wildlife documentaries were constructed representations, whose knowledge value is enhanced by such obvious artificiality. To disclose that the idea for the film came with no specific tree in mind suggests that the tree appearing in the programme stands for all the trees of its kind, the equivalent of a type specimen in natural history. The constructed nature of the wildlife television programme appears as a

---

[28]Anonymous, 1963, 'The major: The story of an oak tree and the creatures that live among its branches', *Radio Times*, 16 May 1963, p. 19.

necessity, and the source of its objectivity. The film here stands as a puri-fied version of nature, a means of experiencing it in a way that no single field trip would allow. Now, if we return to *Unarmed Hunters*, seen along-side Parsons's two early ventures into feature productions—*The Unknown Forest* and *The Major*—it is not a training film anymore but a making-of documentary, disclosing the materials and methods used in producing a new kind of wildlife film, and demonstrating the professionalism at work in Bristol.

Bringing Wildlife Television Closer to Science

This definition of wildlife television-making as a profession, based on the expert handling of film-making technology, participates in a cultural trend which structured public debate in Britain in the first part of the 1960s. In its effort to return to power, the British political left, embodied in the Labour Party, joined forces with part of the scientific establishment to elaborate policies that would align scientists' agenda—increased scientific education and science funding—with those of the Labour Party, whose leadership in the late 1950s believed in the idea of scientifically informed rational government. Meetings between the Labour leaders Hugh Gaitskell and Harold Wilson and such publicly visible scientists as Solly Zuckerman, Patrick Blackett, and Jacob Bronowski laid the basis for the emergence of an ideology that equated technocracy as the rule of professionals deploying their technical expertise for the common good with the notion of social progress (Nye 2004: 158–159; see also Ortolano 2011). Such a visible association with scientists characterised the British Labour Party as the party of progress and modernity. By contrast this rhetorical association painted Harold MacMillan's Conservatives as politically regressive ama-teurs who failed to grasp the value of science and technology for effective government. Anthony Sampson's use of the dichotomy in his 1962 best-seller *Anatomy of Britain*, gives a sense of the penetration of this rhetoric in the public debate:

The hereditary establishment of interlocking families, which still has an infectious social and political influence on the Conservative party, banking and many industries, has lost touch with the new worlds of science, industrial management and technology, and yet tries to apply old amateur ideals into technical worlds where they won't fit. The menace of the British Conservative nexus, it seems to me, lies in the fact that it has retreated into an isolated and defensive amateur world, which cherishes irrelevant aspects of the past and regards the activities of meritocrats and technocrats as a potential menace. (Sampson 1962: 633)

Eileen Molony's criticism of Peter Scott as being 'resistant to innovations on past practice' is another iteration of this rhetoric differentiating progressive professionalism based on an engagement with science and technology from conservative amateurism. The NHU's adoption of this progressivist discourse is indicative of, as much as it reinforces, the construction of science as a rhetorical resource for authority and power in the public realm taking place in early 1960s Britain.

But the NHU's eagerness to build visible links with science and scientists, effectively marginalising amateur naturalists in the process, was also motivated by the increasing public visibility of a new field of scientific inquiry: ethology. Ethologists specialise in observing animals in the field, as opposed to the laboratory, and are concerned with understanding behaviour from the perspective of the theory of evolution. Rooted in the amateur culture of natural history, ethology was the last scientific discipline to grow out of it (Nyhart 1996). Since the 1950s, several publicly visible scientists had been active to raise ethology's public profile. Niko Tinbergen and Konrad Lorenz, who in 1973 would share a Nobel Prize with Karl von Frisch as inventors of the discipline, often featured in television programmes and had been prolific authors of popular books on the topic since the early 1950s. Desmond Morris, one of Tinbergen's early Ph.D. students, was also largely contributing to the spread of the ethology gospel in Britain through his *Zoo Time* programme. At the beginning of the 1960s, Tinbergen had also begun producing films on animal behaviour destined for a large audience, thanks to a grant from the Nature Conservancy. Wildlife broadcasters, as another nascent profession developing from the same amateur culture of natural history and similarly concerned with the study of animal behaviour in the field, could arguably be nervous about the development of a public discourse on the topic originating from the scientific community. Instead of competing, Desmond Hawkins was keen to develop close collaborations.

> Although there are many respectable motives for an interest in wild-life (as well as some disreputable ones) the spirit of scientific enquiry must have pride of place. In handling this subject we expose ourselves to the critical scrutiny of scientists, and their approval is an important endorsement. Moreover, it is their work that throws up the ideas and instances and controversies from which programmes are made. We look to them as contributors, as source material, as consultants and as elite opinion on our efforts. In short, we need their goodwill.[29]

The professional culture of wildlife television making, which the Bristol broadcasters strove for, needed to create its own space between the two pre-existing cultures of wildlife film-making linked either to amateur natural history (represented by Peter Scott) or to big game hunting (represented by the Denises). To foster close associations with scientists for the purpose of producing television programmes was seen as the way to develop a 'professional view of Natural History',[30] one where television could contribute to the production of natural historical knowledge.

Some links between NHU broadcasters and scientists studying animals already existed. Producers such as Eileen Molony and Bruce Campbell regularly attended international conferences on various aspects of zoology such as animal behaviour or ornithology. There they got ideas for programmes and met with some of the scientists who would later contribute to them. International scientific conferences were also a setting where producers could read papers or show some of the NHU's films. For example, Jeffery Boswall gave a paper on the recording of bird songs at the annual meeting of the British Ornithologists' Union in York in March 1961. In September the same year, Eileen Molony attended the

[29] Desmond Hawkins, 1962, 'The BBC Natural History Unit. Report by the head of West Regional Programmes', p. 7. BBCWAC R13/462/1.
[30] Memo: Assistant Head of Talks (General), Television to A.C.P.Tel, 'BBC Natural History Unit', 11 October 1962, p. 1. BBCWAC T31/386.

congress of the Scientific Film Association in Rabat. She took with her special prints of *The Unknown Forest* and *Galapagos* (a compilation of Peter Scott's *Faraway Look* series).[31] However, these relationships were only loose ones. Hawkins was dedicated to strengthening television as a medium and promoting a scientific world view. His ambition was for broadcasters and scientists to collaborate in the creation of a public culture of natural history informed by scientific research and communicated through television. This 'would give the Unit an unequivocal status in the scientific world and would declare the Corporation's intention to make a deliberate creative contribution to science in the film medium'.[32] Between 1961 and 1963, Hawkins explored several avenues going in this direction.

In the first place, when the time came to find a successor to naturalist Bruce Campbell at the head of the NHU, he attempted to appoint a zoologist who would hold at the same time a lectureship in Zoology at the University of Bristol. Campbell was a noted ornithologist with a Ph.D. in Comparative Bird Studies from the University of Edinburgh but no broadcasting experience. To replace him, Hawkins's ideal candidate was Desmond Morris, an Oxford-trained ethologist and seasoned broadcaster from the London Zoo (Chapter 3). Discussions between Morris, the NHU and the University of Bristol reached an agreement before Morris suddenly pulled out. At the last minute he could not resolve himself to leave his position of curator of mammals at the London Zoo, with all the research and potential it offered.[33] Instead, the position went to Nicholas Crocker, a seasoned broadcaster, who had previously been head of the Documentary and Outside Broadcasts Unit at BBC West in Bristol, and who had already been head of the NHU for a year before Bruce Campbell took the position (Chapter 2). For want of strengthening the NHU's standing with scientists, the appointment at least reinforced the NHU's film-making capability and developed its professional standard.

Meanwhile, Hawkins, together with the senior management at the BBC and the University of Bristol, began exploring the possibility of 'setting

---

[31] Bruce Campbell, 1962, 'Report of the Natural History Unit for 1961', p. 9. BBCWAC R13/462/1.

[32] Desmond Hawkins, 1962, 'The BBC Natural History Unit. Report by the head of West Regional Programmes', p. 8. BBCWAC R13/462/1.

[33] Frank Gillard, 1961, 'Natural History Unit: Senior producer post', Memo to DSB, 11 December 1961, BBCWAC R13/462/1.

up some kind of Institute of Animal Behaviour in Bristol [which] would provide our Unit with a badly needed laboratory which would provide the original film increasingly required by *Look* and other series'.[34] This imagined place was to be 'a British Seewiesen, developing animal behaviour studies in the Lorenz manner'.[35] There, under the supervision of the NHU Editor, television producers would work with research scientists on the presentation of the latter's experiments and studies in broadcasting terms. In the same movement, and in no uncertain terms, Hawkins directly refers to ethologists and the necessity for the NHU to collaborate with them: 'if Bristol University, or any other, were inclined to initiate, there would be a strong inducement to the Corporation to examine the possibilities of a partnership'.[36]

> If we are to expand our output of illustrative sequences and behaviour studies we need the accommodation to house and display to cameras the small animals that would be suitable subjects: space in effect for aquaria, vivaria, cages, a lit filming area and some outdoor locations. The distance from Bristol should not exceed twenty miles. If the outdoor setting were sufficiently mixed (meadow, woodland, water, etc.) it would have O.B. [Outside Broadcast] and sound recording possibilities added to its usefulness for filming.[37]

As much as this project could transform the methods of wildlife television production, it also had the potential to transform the way scientists worked, using film-making as a method of inquiry. In both cases, Hawkins's project would have positioned the television institution as an active participant in the production of natural knowledge on the ground of producers' technical expertise in handling the film-making apparatus. In the end, though, Hawkins's vision did not materialise. But as the next chapter will discuss, his effort to align the Bristol unit more visibly with

---

[34] Ibid.
[35] Desmond Hawkins, 1962, 'The BBC Natural History Unit. Report by the head of West Regional Programmes', p. 8. BBCWAC R13/462/1.
[36] Ibid.
[37] Op. cit., p. 9.

professional science was symptomatic of the return to the forefront of public debate in Britain of the cultural politics of broadcasting, whose main catalyst was the work of the Pilkington committee and the publication of its report on the future of broadcasting in June 1962. The main impact of this report was the creation of BBC2, whose launch, in 1964, created an opportunity for David Attenborough to realise part of Hawkins's vision in such programmes as *Life* (fronted by Desmond Morris), and to experiment with an approach to wildlife television informed by close collaborations between scientists and broadcasters.

References

Boon, T. (2015). 'The televising of science is a process of television': Establishing Horizon, 1962–1967. *British Journal for the History of Science, 48*(1), 87–121.

Burns, T. (1977). *The BBC: Public institution and private world.* London: Macmillan.

Gouyon, J.-B. (2016). Science and film-making. *Public Understanding of Science, 25*(1), 17–30.

Nye, M. J. (2004). *Blackett: Physics, war, and politics in the twentieth century.* Cambridge and London: Harvard University Press.

Nyhart, L. K. (1996). Natural history and the "new" biology. In N. Jardine et al. (Eds.), *Cultures of natural history* (pp. 426–443). Cambridge: Cambridge University Press.

Ortolano, G. (2011). *The two cultures controversy: Science, literature and cultural politics in postwar Britain.* Cambridge: Cambridge University Press.

Parsons, C. (1969). The silent watcher. In J. Boswall (Ed.), *Look: A selection from the BBC-TV natural history series* (pp. 13–19). London: British Broadcasting Corporation.

Parsons, C. (1982). *True to nature.* Cambridge: Patrick Stephens.

Sampson, A. (1962). *Anatomy of Britain.* London: Hodder & Stoughton.

Scott, P. (1966). *The eye of the wind.* London: Hodder & Stoughton.

Shapin, S., & Schaffer, S. (1985). *Leviathan and the air-pump.* Princeton: Princeton University Press.

.

6

Showcasing Science, Showcasing Nature on BBC2

On 13 July 1960, the British Government appointed the Pilkington Committee, as it came to be known after the name of its chairman, Sir Harry Pilkington, to consider the future of British broadcasting. One of the questions the committee was required to answer was whether the BBC or the ITA should provide additional service—namely, be given a further channel. In its report which was published on 27 June 1962 and was very dismissive of commercial television, the committee recommended that the BBC should be given an additional television channel; BBC2 was launched in April 1964. The Pilkington Committee had mandated that the BBC use this new channel to increase the broadcasting of educational content, science, in particular.

Until March 1965, the controller of BBC2 was Michael Peacock. Under his oversight the BBC NHU contributed almost nothing to the new channel's schedule, save for a few repeats of the best *Look* programmes, aptly retitled *Look Again*. But in March 1965, David Attenborough was invited to replace Michael Peacock as the head of BBC2. From then on, as Christopher Parsons remembered,

© The Author(s) 2019
J.-B. Gouyon, *BBC Wildlife Documentaries in the Age of Attenborough*,
Palgrave Studies in Science and Popular Culture,
https://doi.org/10.1007/978-3-030-19982-1_6

after Attenborough's appointment (…) things began to look up for us. We were given a brand-new fortnightly series called *Life*, presented by Dr Desmond Morris, and this began in November—just two months after Bristol began receiving BBC2 pictures from the UHF transmitter at Wenvoe in South Wales. (Parsons 1982: 237)

With the creation of BBC2, and the appointment of David Attenborough as its controller in March 1965, wildlife television in Britain took a decisive turn toward scientifically informed coverage of natural history topics, reconfiguring the relationship between nature and humans, filmmaking and science. Until then, two cultures of wildlife television had prevailed, competing to shape the spectacle of nature on British television screen. One originated in the culture of imperial big game hunting. It was best represented in the 1960s by Anglia TV's *Survival* series. An earlier example had been the work of Armand and Michaela Denis. The other, coming from the Victorian culture of amateur natural history, was epitomised in Peter Scott's *Look* series, and defined the output of the BBC's NHU for the first ten years of its existence. Both cultures rested on a firm division between nature and culture, naturalist cameramen priding themselves on their ability to obtain images of animals without intervention. Joining forces with scientists, the NHU successfully established what was effectively a third culture of wildlife television. Celebrating a scientific approach, this culture considered wildlife from an anthropocentric standpoint, defining it in terms of economic value and its usefulness for humanity, as a resource for the production of knowledge and as a mirror held to society, a source of social political order.

The development of ethology, the scientific study of animal behaviour, in the late 1950s and early 1960s, was key to this redefinition of the relationship with nature as presented on the BBC. At the same time as wildlife television broadcasters were breaking away from amateur naturalists and big game hunters, seeking to fashion themselves as professionals, ethologists engaged in redefining wildlife conservation as a scientifically informed project, similarly distancing themselves from earlier approaches inherited from practices of game management and the natural historical aesthetic valuing of nature. The BBC2 programme, *Life in the Animal World* (1965–1968), the series which ethologist and television personality

Desmond Morris fronted for slightly more than two years, encapsulates this development. Offering the spectacle of zoologists publicly debating controversial issues in their field, *Life* forwarded the notion that wildlife television should be the outcome of a collaboration between professional film-makers and scientists. This programme, which turned into an outlet for Morris's 'human ethology', was pivotal in asserting the link between television displays of animal life and social political order.

In its approach to wildlife television, *Life* reversed the relationship between the topic—natural history—and the medium—television. During the single channel era, wildlife television had celebrated the culture of amateur natural history, with programmes front-staging such well-known amateur naturalists as Peter Scott and produced with an audience of amateur naturalists in mind. Television was a means for naturalists to present their pursuit to a broader audience. The brand of wildlife television offered to audiences on BBC2 after 1965 appropriated natural history as a topic to make television programmes. Instead of amateur naturalists, it brought professional broadcasters and film-makers in collaboration with scientists for an audience imagined as less specialised, but also, as will appear, defined as more 'adult'. This shift completed the displacement of expertise from the sphere of amateur natural history to that of institutional broadcasting initiated at the start of the 1960s, as portrayed for example in Christopher Parsons's documentary *Unarmed Hunters*. It enabled wildlife broadcasters on the BBC to create programme formats that whilst re-appropriating key markers of traditional wildlife film-making, such as the travelogue, stood in a renewed relationship to nature.

Life Opens Up Expert Disagreement to Public Scrutiny

The fate of Armand Denis, whose films about exotic wildlife had formed the basis of some of the earliest natural history programmes broadcast in Britain, is emblematic of the mutation wildlife television began undergoing after 1965. Denis's collaboration with the BBC had started in 1954, and for a time, his programmes clashed with the first *Zoo Quest* series, as both he and Attenborough visited similar places, filming similar species.

But in May 1965, Denis's series *On Safari*, overseen in Bristol and broadcast since 1956, ended abruptly. Desmond Hawkins informed Denis:

> I am afraid I have bad news for you. I had my first meeting with Michael Peacock as the new Controller of BBC-1 and he told me he wanted to end the run of *On Safari* after the present series. This is not from immediate dissatisfaction with your latest programmes. It is simply that the general pattern of *On Safari* has had a longer run than almost anything in television and he feels that the format has become hackneyed and old fashioned.[1]

The dismissal felt brutal to Denis. Yet, unbeknownst to him, his work had come to typify, for some time at the BBC, everything from which broadcasters wanted to distance wildlife television. In 1962, Huw Wheldon, the assistant head of Talks, had thus expressed misgivings about the Denises' work when discussing the direction wildlife television should take:

> We, as you know, have always taken the view here that the Armand and Michaela Denis programmes are badly sullied in this connection and that it does neither the Travel and Exploration Unit nor the Natural History Unit much good to have them around.[2]

Not one to be easily dismissed, however, Denis tried, through his agent, to have *On Safari* relocated on BBC2. David Attenborough, who by then had become the channel's controller, would not even consider it. Writing to Denis, he explained that the series did not fit in his project for the broadcasting of natural history on the new channel. On the contrary, *On Safari* represented the type of programme Attenborough wanted to abandon:

> BBC-2 will seek new types of programmes and deal with subjects that have been ignored elsewhere. From this, you will readily appreciate that it would be a denial of our stated objective were we simply to take over programmes that have dropped out of BBC-1's schedules. Thus there is a problem.

---

[1] Desmond Hawkins, personal letter to Armand Denis, 28 May 1965, BBCWAC SW3/20/1.

[2] Assistant Head of Talk, 'BBC Natural History Unit', memo from to Assistant Chief of Programmes, Television, 11 October 1962, BBCWAC T31/386.

When I arrived here, BBC-2 had no Natural History programme whatever, and, as you may imagine, I was anxious that it should have a regular one as soon as possible. But equally we feel it would be wrong to try to produce a carbon-copy of either 'Look' or 'On Safari'.

At the moment, we have scheduled a new magazine dealing with Natural History in general, from a fairly scientific point of view. … What this will be like, and how it will go, we do not yet know, for indeed, the first issue does not go on the air for another few weeks.[3]

The absence of wildlife programmes on BBC2 when Attenborough took responsibility for it provided him with a blank slate. The 'BBC-2 Master Pattern', designed, notably, by his predecessor Michael Peacock, gave him the freedom to make a rupture from previous models of wildlife broadcasting.[4] These guidelines insisted on 'the need to establish a new and different character for BBC-2', which had to be 'real and long-term'. One aim was to convince 'millions rather than thousands' that it was worth buying a 625-line UHF television set, the new technical standard used to broadcast on the new channel. Another was to adapt to 'the changing political climate in the field of educational television'. Applied to wildlife television, this translated to the creation of new programme formats, turning BBC2 into a laboratory where a new cultural space could be carved, one that was structured around a close collaboration between programme makers and scientists and valued film as a means of obtaining knowledge of nature. In a sense, BBC2 became an on-air version of the shared location for which Desmond Hawkins had been making plans in 1961, eventually reshaping wildlife programming on BBC1 as well.

Attenborough, when writing to Armand Denis, had kept his plans for the wildlife output on BBC2 under a veil of relative mystery. However, in an earlier internal correspondence with Aubrey Singer, the head of Outside Broadcast, Features and Science, he had provided more details about what the new natural history programme would be:

---

[3]David Attenborough, personal letter to Armand Denis, 21 October 1965, BBCWAC SW3/20/1.
[4]Michael Peacock, 'The BBC-2 Master Pattern. Note by Chief of Programme BBC-2', 16 October 1963, BBCWAC T62/196/1.

I do not want any natural history programme on BBC-2 to be a twin of 'Look'. It is important that the new programme should have an entirely different basis. The most exciting new line of thought in the natural history world is the scientific examination of animal behaviour. Desmond Morris of the London Zoo is by far the most experienced television practitioner in this field and the least used on BBC. He will be the link man and responsible for editorial content. It is inevitable, therefore—and in my eyes desirable— that behaviour studies should form a very important component of the programme.[5]

Initially titled *Fauna*, the new programme became *Life in the Animal World*, soon abbreviated to *Life*. Recorded in a studio in the style of a magazine show, a typical episode would feature several items including discussions of such topics as the nature of animal aggression, population control in animals and humans, animal communication signals, and the analysis of parent/offspring relationships. To illustrate them, film sequences were projected on a screen in the studio, or animals were brought on the set. The speakers were such luminaries of the 1960s life sciences as conservationist Bernhard Grzimek, geneticist John Paul Scott, primatologist John Russell Napier, and life scientist Miriam Rothschild. On the surface, the format of *Life* was not an innovation, as it shared much in common with *Look*, or even the earlier *Zoo Quest*: studio conversations illustrated with film sequences. What was new, however, was the scientists' role in a programme presented as being about natural history and wildlife, and the way film sequences were put to work. Specially shot by crews from the BBC's NHU (or commissioned externally), they were to support scientists' claims to knowledge and make them visible to audiences. *Life* forwarded the notion of a close association between film and the scientific study of animal behaviour on television. In this new relationship, no room was left for amateur naturalists' expertise.

Attenborough had become interested in the scientific study of animal behaviour when working part-time for the BBC whilst studying for a Ph.D. in Anthropology at the LSE (a project brutally aborted when he was offered his management position at the BBC). During this more

---

[5] David Attenborough to Aubrey Singer, 8 June 1965, BBCWAC T14/2194/1.

academic period, he often attended the research seminar which Desmond Morris was conducting at the London Zoo. As Desmond Morris recalls:

> At the time of my zoo seminars (1963–1965) David Attenborough had for a while left TV and was studying Anthropology at London University. He asked if he could sit in on my zoo seminars and did so frequently. When he was asked to return to the BBC to start the new channel BBC2, he wanted to introduce some new kinds of TV programmes and, remembering the seminars, must have got in touch with Bristol and asked the Head of the Natural History Unit there (Crocker) to contact me and see if I would do a series for them in addition to my ZOOTIME series for TV. After meetings with Nicky Crocker I agreed and was excited by the new challenge. And many of the people who were guests on the LIFE series were people who were attending my seminars at the zoo.[6]

This early discussion of the programme between Crocker and Morris took place in late October 1964, over lunch at the London Zoo. As a magazine, it would contain topical natural history subjects, but also 'non-topical items, which would not only have an "on-the-shelf-value"' but could be stored in the NHU's library of footage to be reused at a later date. However, when at the end of November Crocker presented the idea to Michael Peacock, still the controller of BBC2 at that time, Peacock turned it down.[7] He was among those at the BBC who thought that wildlife should be approached from a scientific perspective, and that the BBC science department in London was better positioned to achieve this than the Bristol unit, whose approach to natural history topics many in London thought belonged to light entertainment and children's television.

One of Peacock's flagship projects for BBC2 had been to launch the science programme *Horizon* (Boon 2015). And from his perspective, what Crocker proposed for *Life* was redundant with *Horizon*. In their search for ideas for *Horizon*, the broadcasters responsible for the new series—foremost among whom was Philip Daly—had turned to natural history as a

---

[6] Desmond Morris, e-mail to author, 17 May 2014.
[7] Nicholas Crocker, personal Letter to Desmond Morris, 9 November 1964, p. 1. BBCWAC WE8/430/1.

source of topics. For example, invited to attend the symposium *The Natural History of Aggression*, featuring speakers like Konrad Lorenz, Daly had started planning for a programme on this theme.[8] It did not materialise, but the second episode of *Horizon*, *Pesticides and Posterity*[9] examined the effects of insecticides on wildlife, featuring contrasting interviews with naturalists, chemists, conservationists and zoologists. Peacock had suggested the idea for the programme, which epitomised the scientific approach to natural history he and others advocated at the BBC. When Crocker came to him with his programme idea, Peacock could only turn it down, especially as he had just ruled that *Horizon* should be broadcast fortnightly rather than monthly.

But Peacock left BBC2 in February 1965 to take charge of BBC1. One of Attenborough's first decisions after he had returned to the BBC to replace him, was to formally accept 'a natural history magazine' from the NHU to be broadcast on BBC2. Aubrey Singer, the head of Science and Features and the man in charge of *Horizon*, challenged the decision on the grounds that his department was already responsible for 'Britain's leading science magazine'.[10] As such, Singer argued, *Horizon* should have priority when it came to producing programmes on natural history topics for the new channel. In a straightforward reply, Attenborough ring-fenced the NHU's territory on BBC2.

> There is a great body of natural history expertise and experience in Bristol and a large natural history film library which is known to the people there and which is under-used. Furthermore, natural history films are automatically sent to West Region [Bristol] for consideration as material for 'Look'. Often there is not enough in any one offering to provide a full 'Look' programme, but a solid kernel of exciting material lasting perhaps five or six minutes would be excellent for this new programme.
>
> …

---

[8]Anthony Storr, personal letter to Philip Daly, 15 November 1963, BBCWAC T14/1810/1.

[9]Broadcast 30 May 1964.

[10]Head of Outside Broadcasts Feature and Science Programmes, Television to C.BBC-2, 'Science coverage', 3 June 1965, BBCWAC T14/2,194/1.

I hope that you may feel that with all the world of science at your disposal it will not be difficult to avoid treating the sort of subjects in 'Horizon' that obviously closely fit the natural history magazine's brief. I would ask, therefore, that when the new programme gets under way, you establish a liaison with West Region on this subject. I hope that difficulties and disputes will not occur, as I am sure you do. If they do arise, I will have to adjudicate.[11]

In this note to Singer, Attenborough makes three decisive moves, setting the path for the next generation of wildlife television in Britain. In the first instance, he asserts the validity of the claim wildlife broadcasters had been putting forth with films such as *Unarmed Hunters*: their expertise in wildlife film-making should be seen more broadly as a new form of natural history expertise in which wildlife television programmes are framed as outcomes of the pursuit of natural historical knowledge. Secondly, Attenborough acknowledges that thanks to its connection with Peter Scott through *Look*, the NHU commands a privileged position in the British networks of natural history. In 1965, the Unit achieved the status of what sociologist of science Bruno Latour called a 'centre of calculation' (Latour 1987: 232), a place generating stable, accepted depictions of the world out of a flow of incoming raw data. Lastly, the note shows Attenborough's awareness of the symbolic capital *Look* had accumulated over the years and the necessity to associate it with any renewed approach to wildlife television if it were to succeed with audiences. Although he conceived of *Look* as an out-of-date model of wildlife programming, he was keen to retain the trust, prestige and recognition associated with it and obtained in the realm of amateur natural history, to re-invest it in the next generation of wildlife television programmes, defined as a branch of science television. The ambition was not to eradicate *Look* but to build upon what this series had achieved in placing natural history on television and helping define the medium as a partner in this knowledge-making endeavour, to try and similarly place zoology on television and define the medium as a partner in the production of zoological knowledge.

---

[11]David Attenborough to Aubrey Singer, 8 June 1965, BBCWAC T14/2194/1.

With the arrival of Attenborough in a managerial position at Broadcasting House, Bristol felt that they had 'a friend in London' (Parson 1982: 237).[12] But equally, Attenborough had very strong views on the form wildlife television should take and was keen to use his position to implement them. His key idea was 'to make natural history programming more scientific'.[13] To him it was 'important' to move away from programmes 'simply … showing the beauties of nature'. Instead, building 'on the widespread public appreciation of natural history', television could invite audiences to examine 'in a serious and critical way new trends and ideas in zoology'.[14] Yet, with this approach, Attenborough was positioning wildlife television in the territory of *Horizon*, a programme similarly intended to 'reflect the current trends in scientific thinking'.[15] Aubrey Singer, himself 'a highly competitive individual' eager to 'gain supremacy over other departments in science broadcasting' (Boon 2015: 91), would not be easily persuaded to enact others' plans. By contrast, the NHU which had felt left out from the new channel, with no programme commissioned for BBC2 thus far, welcomed Attenborough's attention. As Desmond Morris remembers, 'David, as Controller, commissioned the series and asked me to include discussions about zoological controversies. Then he left it up to Crocker and [Ronald] Webster [the producer for the series in Bristol] to discuss specific items and subjects with me'.[16]

With *Life*, Attenborough temporarily achieved, and at the scale of a programme, what had been Desmond Hawkins's ambition when he'd attempted to hire Desmond Morris as head of the NHU. It created a professional context for television producers to work with the ethologist and, under his guidance, produce film representations of research scientists' experiments and studies, in television terms. The producer put in charge of the programme, Ronald Webster, was a seasoned broadcaster. Under him were three debuting production assistants, Richard Brock (1938–),

---

[12]Sheila Fullom, personal communication to author.

[13]Desmond Morris, e-mail to author, 16 May 2014.

[14]David Attenborough to Solly Zuckerman, 8 June 1966, BBCWAC TVART3—Dr. Desmond Morris—File 2—1963–1970.

[15]Daly to Singer, 'Horizon magazine programmes', 5 March 1963, BBCWACT14/3,316/1 (quoted in Boon 2015: 97).

[16]Desmond Morris, e-mail to author, 16 May 2014.

John Sparks (1939–) and Barry Paine (1937–2011), each of whom became, in the following years, a key actor of wildlife television. Brock and Sparks both went on to produce episodes of *Life on Earth* (1979). Paine produced many episodes of the series *The World About Us* (1967–1986). Through their work on *Life* they learned the ropes of television broadcasting and got into the routine of working with scientists to produce film sequences about the latters' research to illustrate sometimes heated discussions held in a studio. As Barry Paine remembers,

> This was broadcasting natural history for big boys, this was the sort of feeling that was going on. We weren't going to dumb it down, we were going to jolly well say it how it is and these are the people that are doing the work, people looking at animals in the wild and we were going to have them in the studio and we were going to have discussions on pretty esoteric topics like aggression.[17]

The first episode of *Life* aired on 14 November 1965, a Sunday, in the evening. The topic, cats and dogs, was one likely to appeal to British audiences, a nation of pet keepers. The line-up included such international figures from the community of ethologists as John Paul Scott, from the Jackson Laboratory in Maine, and Rudolf Schenkel, from the University of Basel. The guests presented their research on the origins of domestic dogs and cats, and their influences on human life. Billing the programme in the *Radio Times*, producer Ronald Webster declared it to be a means of sharing 'new ideas on animal life' with viewers whose interest in the subject had previously been expressed 'in a steadfast loyalty to the programme output of the BBC's Natural History Unit'.[18]

From the outset, *Life* was thus targeted at viewers used to such programmes as *Look*, but with the avowed aim of shepherding them away from that type of approach. Still describing the series to *Radio Times* readers, Webster characterised the new series as a progression from earlier wildlife television programmes. *Life* was to offer viewers 'a sample of scientific thought' and, if it asked questions, did not 'necessarily propose to

---

[17]Barry Paine, Oral history interview, 31 January 2001, Wildscreen.
[18]Ronald Webster, 'Life', *Radio Times*, 11 November 1965, p. 15.

provide definitive answers—because there may be none yet'. Whereas earlier wildlife television programmes such as *Look* had encouraged viewers to 'sit back and bask in the beauties of nature', and had tended 'to lay down the law about the "known facts" of animal life', *Life* would 'show zoologists thinking, arguing, and doubting new ideas about the animal world'.[19] True to Attenborough's ambition, *Life* was an invitation to encounter science in the making, cast as a questioning activity. From this standpoint, film provided the means for viewers to witness and share in the process of knowledge production. Watching the film sequences illustrating the discussions taking place in the studio, they could engage in some form of questioning of their own and thus participate in the knowledge production activity from a distance—in their living room. Simultaneously, as a public forum to discuss their ideas, *Life* offered scientists a space to solicit support from viewers for their claims to expertise, notably on topics considered the preserve of naturalists. This is exemplified in *Life's* approach to wildlife conservation, a topic which naturalists had owned in the public sphere in the post-war decades but which scientists, notably ecologists and ethologists, were starting to appropriate in the mid-1960s.

Wildlife Conservation as a Scientific and Televisual Project

The conservation of wildlife had been the topic of several *Look* episodes over the years. In 1955, *Wildlife in Trust* featured Mervyn Cowie, director of the Royal National Parks of Kenya, in a discussion with Scott on maintaining the balance of nature in African National Parks. In 1961, in a special episode, *L for Lion*, Scott announced the launch earlier that year of the WWF, whilst in 1964, in *Conservation in Action*, he presented the WWF's work across the world. In all these episodes, the key actors of wildlife conservation on the ground appeared to be naturalists, activists, members of clubs and societies, with institutions—foremost amongst which was the WWF, in later episodes—providing the direction of travel. To justify the conservation project, Scott resorted to moral and aesthetic arguments,

---

[19] Ronald Webster, 'Chimpanzee—Fallen man?', *Radio Times*, 9 December 1965, p. 14.

both rooted in the culture of amateur natural history. From this perspective, wildlife parks and natural reserves were the place of conservation at the same time as they were areas of individual regeneration where people could come to be 'refreshed and recreated', as Scott explained in *Conservation in Action*. Wildlife conservation was necessary, and the work of such bodies as the WWF justified, to ensure that future generations could still enjoy the spectacle of lions or other endangered charismatic megafauna in the wild. As a moral project, wildlife conservation was also a collective one, in which programmes such as *Look* were meant to enrol viewers.

By contrast, in line with Attenborough's declared ambition that *Life* should not be 'another natural history programme showing the beauties of nature',[20] the series approached the question of conservation from a decidedly pragmatic and rational perspective. For example, instead of celebrating the success of a piece of legislation partly originating in conservationist lobbying (the 1963 Deer Act), the 1966 episode *Wild Animals at Large* discussed the effect of deer overpopulation in Britain, evoking a 'silent fifth column' to question the place of 'so many deer in an urban society'.[21] Departing from the moral and aesthetic line of argument that earlier advocates of wildlife conservation had put forward, *Life* promoted a pragmatic view of conservation as scientific management. In this episode on deer, Morris brought in, as discussants, representatives of the Forest Commission and the Nature Conservancy, two organisms responsible for land and natural resources management in the UK. Inevitably, the notion of management raised the question of who should be in charge. The programme suggested that the experts to be entrusted with the project had to be scientists.

Three *Life* programmes broadcast over the summer of 1967 extolled this view of conservation as a scientific and economic project. These episodes, where Attenborough acted as a presenter, focused on the scientific work conducted at the Serengeti Research Institute (SRI) in the Serengeti National Park in Tanzania. Interviews with John Owen, the

---

[20] David Attenborough to Solly Zuckerman, 8 June 1966, BBCWAC TVART3—Dr Desmond Morris—File 2—1963–1970.

[21] Ronald Webster, 'Life', *Radio Times*, 18 August 1966, pp. 15, 13.

director of the Tanzanian National Parks, with the SRI's scientific director, Hugh Lamprey, and with Niko Tinbergen, as chairman of the SRI's
scientific committee, repeatedly put forward the key message that scientific research, as opposed to game keeping and natural history, was the
necessary foundation of wildlife conservation, defined as an economically
valuable practice, bringing wealth to the country. National parks in Africa,
such as the Serengeti, and their wildlife, are presented in these interviews
as the last remaining strongholds of primeval nature which are also key
assets for the tourist trade. There, scientists can access complete ecological
knowledge and in turn put this knowledge to work to maintain wildlife
parks in a state that will keep meeting tourists' expectations and help bring
foreign currency to African states. Owen, the National Parks director, casts
scientists as experts—contrasting them with witchcraft men—providing
wildlife park managers with the tools and knowledge they need 'to keep
the parks into balance'. In the words of Niko Tinbergen, 'Scientists are a
vital part of the machinery of the tourist trade. Their work has undoubtedly hard cash value'. Lamprey holds a similar discourse, asserting the
necessity of scientific knowledge for the proper management of wildlife
parks, conceived of as epistemic and economic assets. Attenborough, for
his part, in quite a surprising swipe at naturalists, warns viewers about
'how much caution we should use when we listen to those big sweeping
generalisations that naturalists are prone to make about the way rhinoceros
or lots of other animals behave',[22] and suggests that more trust should be
placed in scientists' much more nuanced and accurate accounts of the life
of animals.

In these three episodes, conservationists, ethologists, and wildlife broadcasters are seen joining forces, seeking to elicit support from the larger
public for their respective pursuits, mutually reinforcing each other in the
public sphere through their joint participation in the same televised spectacle of professional conservation in action, in the service of Western-led
post-colonial development. All three groups, at the same time but for different reasons, were eager to distance their institutions and their practices
from the two cultures of amateur natural history and imperial big game
hunting, which had ruled over wildlife film-making, wildlife conservation

---

[22]BBC, 'Life #42: Tools for research', first broadcast 11 July 1967.

and the production of knowledge about wildlife since the last decade of the nineteenth century. In the wake of the independence movement, conservationists had cast themselves as facilitators of the development of the new nations, in the hope that they would thus be able to see the parks through such a period of political turmoil. But calls for an Africanization of the parks' administration had become increasingly difficult to ignore for the Western conservation elite effectively controlling the parks through a network inherited from the former colonial administration. To be seen still standing in an association with such representatives of the old structures of power in the parks as game wardens, would have weakened conservationists' stance. Scientists, cast as politically neutral providers of expert knowledge, were less problematic in this respect. Having them, instead of game wardens, manage the parks increased conservationists' prospects of remaining in control.

For ethologists, the association with the conservationist project brought funding and new objects of study, as well as a sense of social utility, all of which could help consolidate the new discipline. Having been just appointed chairman of the scientific committee of the newly formed SRI, Tinbergen used his participation in this programme presenting the scientific work conducted there to assert the Institute's legitimacy. His position also provided Tinbergen with a tribune to make the case for an approach to wildlife conservation informed by ethology, thereby connecting his new discipline with mainstream concerns.

> It is one of the basic points of our policy with the Serengeti Research Institute that we must remain acceptable and even welcome in Tanzania and that we can remain so (and so continue our work), if we are seen to be useful to them. I have to bring out how fortunate it is that the interests of terrestrial ecology and of Tanzania coincide. This key point must be made clear, so that both sides will gradually accept it as part of their own creed.[23]

In Niko Tinbergen, the NHU had found a willing participant in their project to renew wildlife television. By offering him an opportunity to present, unchallenged, his views on the value of the presence of Western

---

[23]Niko Tinbergen, personal letter to Richard Brock, 16 March 1967, BBCWAC WE8/600/1.

scientists in the recently decolonised African country's national parks, and by sending no less than its star wildlife presenter to the Serengeti, David Attenborough, the BBC was putting its symbolic power at the service of Tinbergen's project. Conversely, for the BBC to be seen performing a visible association with scientists in the Serengeti—a location previously associated with naturalists' aesthetic and moral ideal, and imperial big game hunters' socially stratified understanding of the access to wild nature—was instrumental in distancing wildlife television from the two cultural repertoires which had previously dominated it. It created a space for a third culture of wildlife television to emerge, one related to the science of animal behaviour. But forwarding this pragmatic, rational and scientifically informed view of conservation was self-serving, and these three episodes of *Life* only presented a partial view of scientific research in the Park.

In 1967, for scientists to be placed at the forefront of the conservation project in Africa still was a very recent development which owed much to the geopolitics of decolonisation. Scientific research in the Serengeti had been formally established in 1961, with the opening of the Serengeti Research Project (SRP), a small research facility hosting two scientists working on the behaviour and ecology of two populations of large herbivores: zebras and wildebeests (Sinclair 2012). The main principles of ethology, the study of animal behaviour in the wild, had been formalised in the 1950s (Burkhardt 2005). Ecology was an older discipline, developed in the last decades of the nineteenth century. But it was not much more organised than ethology, and in any case, its association with conservation was still, in the 1960s, quite new (Bocking 1997). Owing mostly to its origins in the cultural space of big game hunting and game keeping (MacKenzie 1988), the main agents of conservation in the field up until then had been game wardens. Their missions in wildlife parks was to monitor animal populations and prevent poaching, or any form of unauthorised access to the game. Through their activity, they usually developed an important natural historical knowledge, which they put to use for tracking animals when accompanying such visitors as elite big game hunters coming to exert their right to collect trophies. To place scientists in the role of experts in the management of wildlife parks was thus to displace the authority of game wardens, who, of course, did not welcome such a development.

Myles Turner, game warden in the Serengeti from 1956 to 1972, experienced the arrival of scientists in his territory. He provides an alternative account which enables us to understand how this change in the politics of wildlife conservation relates to the history of wildlife television (Turner 1987). Until 1964, the presence of only two scientists at the SRP did not undermine the prevalence of the game-keeping approach to conservation in the Park. But in 1964, after the SRP changed its name to SRI, game keeping lost its position to science as the main informing principle of conservation in the Park.

> Word must have got back to the seats of learning about the opportunities in the Serengeti, because from then on we were inundated with scientists of many nationalities … and a determined smash-and-grab raid for PhDs was started by youngsters who regarded the Serengeti and its animals as a vast natural laboratory to be looted at will. (Turner 1987: 157)

Eventually, in 1966, larger research facilities were built next to the Park, in Senora, with funds from a consortium of Western philanthropic foundations.[24] Hosting up to twenty scientists, this institute, whose director was Hugh Lamprey, was the setting for the three *Life* episodes presenting the research conducted in the Serengeti. To Turner, 'the arrogance of some of these scientists—with the ink hardly dry on their graduation papers—was unbelievable'. All were 'sublimely confident that they had the answers to all East Africa's game problem' (Turner 1987: 157–158). At first, the relationship between wildlife conservation old timers and these new arrivals was one of toleration, despite the scientists' 'eccentric life style, speeding around the Park with their long hair and odd clothes' (p. 158). However, things turned sour when these eccentric scientists began requesting permission to kill animals in the Park for scientific purposes. The question of who was allowed to kill, what kind of animal and for what purpose, was consubstantial to the essence of the wildlife park in Africa. Game wardens'

---

[24]The main grant came from the Fritz Thyssen Stiftung. Additional buildings were provided by the Caesar Kleburg Foundation. Operating costs were provided by the Ford Foundation, Canadian International Development Agency (CIDA), and Tanzania National Parks.

raison d'être was to protect the animals living in the Park only insofar as it enabled the select few who enjoyed such a privilege to be able to kill exotic big game (MacKenzie 1988; Ritvo 1987). From the perspective of game wardens, the story of the scientists' arrival in the game reserve is one of resentment and resistance when faced with what they perceived as a take-over by scientists of wildlife conservation, and a redefinition of its aims and practices in ways that made game keepers irrelevant.

During his stay, David Attenborough interviewed Myles Turner on the topic of wildebeest migration, one of the major natural phenomena taking place in the Serengeti. Only a few months before the interview, game wardens had been informed that the newly formed SRI was taking over all game work in the Park (Turner 1987: 71). Yet, creating for the viewers the impression that all is fine in the Garden of Eden, the interview, as it was broadcast, did not come close to discussing the changing role of game wardens following the arrival of scientists in the Park, therefore concealing the contested nature of the conservation project. Likewise, the portrayal of scientists' activities in the Serengeti did not touch upon the debate in which they were opposed to game wardens over the killing of animals. For instance, a sequence depicting work on the ecology of the African buffalo shows buffaloes anaesthetized, tagged and equipped with a radio collar to track their movements around the reserve. The sequence was filmed as a hunting one, from inside a 4 × 4, at full speed, chasing a buffalo. A researcher is shown aiming a compressed-air rifle at his animal target and firing the dart, but at no point is it mentioned that scientists' marksmanship was not just used to anaesthetize animals and that some of the research conducted on buffaloes at the SRI also involved killing them.

The three *Life* programmes constructed a view of the situation on the ground which asserted the legitimacy of scientists' presence and role. It also helped cement a relationship of mutual dependence between broadcasters and scientists, which would become helpful in the long run for television to return to the parks regularly. More broadly, *Life* brought on the air the spectacle of a tight entanglement of television making with scientific research, of a close collaboration between broadcasters and scientific researchers. The series epitomised the approach to televising natural history favoured on BBC2 under the controllership of David Attenborough. It consolidated the notion that wildlife television in Britain should centre on scientific zoology rather than amateur natural history.

The Human Animal: Discovering Clues About Human Nature in the Study of Animals on Television

Life was propelled by the optimistic belief, widespread among progressists in Britain in the 1960s, that boundless possibilities became available to those joining forces with science. This belief that a new, rational and progressive social order could be forged in the white heat of science and technology found an expression in the way *Life* put the spectacle of animals to work to forward a specific social order. One outcome of the programme was a new field of scientific inquiry, human ethology, whose main proponent was Desmond Morris, who summarised it in his 1967 programmatic best-seller *The Naked Ape* (Morris 1967). Morris developed his ideas about human ethology and the value of studying animal behaviour to discover the rules governing human behaviour whilst working on the series. By his own account, 'It was during the *Life* series that the idea of *The Naked Ape* book finally took shape in my mind'.[25] The book, destined to gain a broad readership, famously encapsulates an attempt at applying the conceptual tools of ethology to the study of human behaviour. The fundamental postulate was that humans, as animals, are liable to the same kind of enquiry as other mammals, or fish, birds, and insects. And as much as *Life* was key to enabling Morris to form the ideas contained in the book, the television programme was instrumental in preparing viewers to receive them positively.

Morris's continuous exploration, in his television series, of the themes he would eventually discuss in *The Naked Ape* had made them part of the lived experience of the show's audiences. A succession of programmes helped turn the claims presented in the book into mainstream ones, making them acceptable even before they had been published. To some extent,

---

[25]Desmond Morris, e-mail to author, 19 May 2014.

the process of peer scrutiny had been performed on the television screen in *Life*, with the viewers as witnesses. *The Naked Ape* can thus be taken as synthesising the approach the TV programme took when discussing animal behaviour in relation to humans. In the two-year run of the series, several programmes attempted to find explanations of human behaviours in animals. For example, the third programme, broadcast on 12 December 1965, focused on chimpanzees, and among the subjects discussed in relation to the ape was that of aggression in animals. Programme participants then attempted to extrapolate from the case of the chimpanzee to explain human aggression. This theme was further explored in a programme broadcast on 6 January 1966, under the title *The Roots of Aggression*, bringing together in the studio zoologists, ethologists and psychiatrists. Similarly, an episode broadcast on 7 March 1967 featured Robert Ardrey, an American playwright who had developed an interest in human origins and had published *The Territorial Imperative* in 1966 (Milam 2019). In his study, Ardrey argued that evidence found in the animal world enabled the understanding of the human instinct 'to defend private property or national boundaries'.[26] The studio discussion involved social anthropologist Robin Fox, specialised in studies of aggression and human behaviour; ethologist Michael Cullen, one of Morris's former colleagues at Oxford; and geologist Sonia Cole, famous for her studies of East African prehistory. *Life* not only worked as a vehicle for Morris's ideas about human ethology but also contributed to establishing the notion that through their taking part in televised spectacles of animal life, scientists had something to contribute to shaping the social political order. Finally, as Morris's *The Naked Ape* hit the bookstores on 12 October 1967, a full episode of *Life* was dedicated to discussing it on the 31st. *The Naked Ape* quickly became a sensation both in the UK and abroad. *Life* had helped lay the ground for this success.

Both *Life* and *The Naked Ape* renewed the meaning conferred on the on-screen spectacle of animal life as exemplary of what human existence should look like in Western capitalist industrial societies. *Life* had been billed as 'a challenge to our accepted ideas about the world of the animal. It reminds us that we are part of that world and that the study of it

---

[26] Robert Ardrey, 'Robert Ardrey', *Radio Times*, 2 March 1967, p. 27

can be an introspective experience.'[27] Morris introduced *The Naked Ape* with an invitation to readers 'to contemplate their animal selves' (Morris 1967: 12), presenting the content of the book as an alternative to both psychoanalysis and anthropology. Anthropology, he said, failed to instruct us about 'the typical behaviour of typical naked apes' (1967: 10) because of anthropologists' focus on 'cultural back-waters so atypical and unsuccessful that they are nearly extinct'.

> They then returned with startling facts about the bizarre mating customs, strange kinship systems, or weird ritual procedures of these tribes, and used this material as though it were of central importance to the behaviour of our species as a whole. The work done by these investigators was, of course, extremely interesting and most valuable in showing us what can happen when a group of naked apes becomes side-tracked into a cultural blind alley. It revealed just how far from the normal our behaviour patterns can stray without a complete social collapse. (Morris 1967: 10)

Far from instructing us about the essence of human nature, anthropological knowledge, as Morris saw it, described unnatural patterns of behaviour that interfered with humans' ingrained drive to progress, as successfully exemplified in the Western industrial cultural sphere. Anthropological knowledge was therefore not helpful as a foundation to construct a model for 'our behaviour as a species'. Morris's argument rested on disputing 'old-style' anthropologists' claim that their 'technologically simple tribal groups' were nearer to 'the heart of the matter than the members of advanced civilizations'. To him, these 'simple' tribes are not primitive but 'stultified', for the human species is essentially an exploratory one, and so any social group that has 'failed to advance has in some sense failed, "gone wrong"' (Morris 1967: 10). By contrast, Morris claimed that the biological study of human behaviour was a much sounder approach, as it rests upon examining 'the common behaviour patterns that are shared by all the ordinary, successful members of the major cultures—the mainstream specimens who together represent the vast majority' (Morris 1967: 12). In this comparison between anthropology and human ethology, patterns of behaviour were explicitly linked to social-political order. Likewise, in

---

[27] Ronald Webster, 'Life', *Radio Times*, 11 November 1965, p. 15

his comparison of human ethology with psychiatry and psychoanalysis, Morris insisted that although mostly preoccupied with individuals from industrial Western societies, what he calls 'mainstream specimens', the latter two disciplines could not be relied upon for discovering the typical patterns of human behaviour.

> The individuals on which they have based their pronouncements are, despite their mainstream background, inevitably aberrant or failed specimens in some respect. If they were healthy, successful and therefore typical individuals, they would not have had to seek psychiatric aid and would not have contributed to the psychiatrists' store of information. (Morris 1967: 11)

With the *Life* series, wildlife broadcasters self-consciously put wildlife television to work as a means of naturalising a social political order premised on the norms and values of Western industrialised post-war capitalist society, enrolling animals in this project. Morris's account does not so much contradistinguish human ethology from other modes of enquiry into the rules governing human nature as it defines the culture and the kind of individuals considered to be most representative of humanity, and worthy of scientific attention. Zoological studies of animal behaviour of the kind carried out by ethologists played a prominent role in supporting the argument presented in *The Naked Ape*, together with information from palaeontology, and field observations of humans in Western industrialised cultures. Sharing the book's focus on animal behaviour studies, *Life* similarly gave wide circulation to the view that such studies enabled the discovery of the actual norms of 'natural' human behaviour. According to human ethology as Morris defines it in *The Naked Ape*, Western industrial capitalist culture and the affluent, white, middle-class individual are the embodiment of such norms. Conversely, human ethology stands as the objective, scientific enterprise naturalising these norms and guaranteeing their veracity. The reliance of human ethology on studies of animal behaviour conferred on televised displays of animal life the status of evidence which could form the basis of viewers' agreement on the claims put forward in the human ethology project. It also turned such displays into vehicles for the norms instituted through the human ethology discourse. *Life* was of its time. It successfully ran from 1965 to 1968, benefiting from and participating in the cultural climate of the sixties, characterised by three features or 'waves' (Agar 2008). One was the opening up of expert disagreement to public scrutiny. Another was the emergence of social arrangements—institutions or audiences—to host such disagreement and exert scrutiny. The third one was an appetite for self-examination and analysis. Yet, this latter aspect is usually associated with the counter-culture, itself an attempt at shaking off diverse forms of the oppression of the self, one of which had been identified as the myth of the objectivity of science: 'At every level of human experience, would-be scientists come forward to endorse the myth of objective consciousness, thus certifying themselves as experts. And because they know and we do not, we yield to their guidance' (Roszak 1969: 209). In this respect, *Life*, and the political project it carried, was reactionary, for it militantly asserted that far from oppressing the self, objective science could reveal it and liberate it.

'Colour Television Is Natural Television': Using Wildlife to Sell Television Sets

For a good two years, *Life* concentrated the efforts of the NHU in Bristol and the approach to wildlife television developed in relation to this programme diffused in the Unit's other output. This is notably evident in the changes that took place from 1966 to 1968 in *Look*, the main wildlife series in the BBC1 schedule, with an increase in the use of films, as opposed to studio discussions, and in the presence of ethologists, displacing amateur naturalists. Jeffery Boswall, the producer of *Look*, announcing the 1966 season in *Radio Times*, gave readers 'a foretaste of the films to come', emphasising the film component of the series. Amongst the highlights was the first episode of the season, *The Private Life of the Kingfisher*, shot by Ronald and Rosemary Eastman, which Boswall billed 'one of the finest wildlife motion pictures ever made in Britain'. In the vein of Heinz Sielmann's *Woodpeckers* (1955), they had managed to obtain shots of birds in their nest inside a riverbank. *The Private Life of the Kingfisher* asserted the power of film-making to reveal hidden truths of nature, as did the Italian film *Flowers without Time* (by Alberto Ancilotto and Fernando Armati), also featured in this season. The 'first *Look* ever without animals', it 'revealed through the technique of fast-motion filming the remarkable movements of plants: tendrils turn-buckling, climbers climbing, buds bursting'.[28] In the 1967 *Look* series, broadcast the year when the BBC ended Peter Scott's contract, the focus on films remained and several episodes presented the work of ethologists, be they Niko Tinbergen (e.g. *Beachcombers*, 22 January 1967), Desmond Morris, in a programme on the panda (9 June 1967), or Jane Goodall, in the National Geographic film *Miss Goodall and the Wild Chimpanzees*, broadcast over two weeks (16 and 23 June 1967). These films, presented as *Look* episodes, exposed BBC1 viewers to ideas about animals similar to those debated in the BBC2 series *Life*, leading to a progressive homogenisation of the outlook on animals and their relation to humans of the BBC's wildlife output. These *Look* episodes also contributed to displacing, within one of their televisual strongholds, amateur naturalists' expertise in favour of that of scientists studying animal behaviour, presented as 'professional naturalists'. Ethologists thus came to stand as wildlife film-makers' partners within a professionalized natural history encompassing both film-making and ethology.

In 1968, for the first season of *Look* not involving Peter Scott, all episodes were film-based ones. Overall, this evolution was in line with the trend Christopher Parsons had started with films like *The Unknown Forest* and *The Major* (Chapter 5). It also matched the dominating feeling at the NHU that film was 'where the future's going to lie', as Nicholas Crocker confided to Parsons at the end of the 1960s.[29] To replace *Look* in the schedule, Boswall produced from 1970 onward an exclusively film-based series, *Private Lives*, broadcast on BBC1. The concept had originated in the immense success of the Eastmans' film on the kingfisher. By the end of the 1960s, the approach to wildlife television Attenborough favoured as controller of BBC2 had thus spread on the two BBC channels. Film-based programmes had become the norm in wildlife television as opposed to live studio programmes. Professional wildlife film-makers collaborating with ethologists had become trustworthy sources of authoritative visual accounts of animal life, displacing amateur naturalist cameramen in the field.

[28]Jeffery Boswall, 'Look', *Radio Times*, 28 April 1966, p. 37.
[29]Christopher Parsons, Oral history interview, 26 July 2001, Wildscreen.

When *Life in the Animal World* ended in March 1968 because its pre-senter, Desmond Morris, had left the UK, the only regular natural history slot in the BBC2 schedule was another Attenborough brainchild, *The World About Us* series. Jointly produced by the NHU in Bristol and the Travel and Exploration Unit in London, it was a film-only series, each episode lasting fifty minutes and not relying on a presenter in a studio. *The World About Us* (*TWAU*) marked a rupture from the conventions of wildlife television programming. Until then, wildlife television had been dominated by presenter-led programmes, with episodes no longer than half an hour. Besides operating a change in programme format, *TWAU* was also intended to use natural history topics as advertisements for a new development in television broadcasting: colour television. Advertised as 'a series of films from all over the world about our astonishing planet and the creatures that live on it', *TWAU* started on 3 December 1967 with Haroun Tazieff's film *Volcano*.

The beginning of *TWAU* coincided with the moment when trans-mission on BBC2 would become entirely in colour, a project presented in public as Attenborough's. As the special section announcing colour television in the *Radio Times* explained: 'Apart from the boffins in the back-room—the engineers—who made the whole thing possible, the man mainly responsible for it all is David Attenborough'.[30] The key message Attenborough wanted transmitted to readers was that his 'main concern is the viewing end'. Viewers had a role to play in making this transition a success: 'You can have the best equipment and the highest transmission in the world … and yet not get the full benefit at home. As in a high-performance car, the machinery can be the finest there is, but the rest is up to the driver.'[31] In other words, viewers had to buy new, colour-enabled sets (an advertisement for Murphy, a maker of colour TV sets featured in the same central section of the *Radio Times* as the interview with Attenbor-ough). Two programmes were brought forth by the BBC as advertisements for colour television: *The Black and White Minstrel Show* and *TWAU*. In charge of *TWAU* in Bristol, Christopher Parsons remembered as much:

[30]Orbiter, 'Round & about. Starting at the centre', *Radio Times*, 30 November 1967, n.p.
[31]Ibid.

I can remember seeing Radio Rentals, saying 'Rent a set and see The World About Us and The Black and White Minstrel Show'. I mean, they were the two things that were really helping to get people interested in colour.[32]

Colour television, Attenborough had explained during the press conference launching the transition to colour broadcasting in December 1967, is television in its 'natural state'.[33] 'Colour', he added for the benefit of readers of a supplement on colour TV in *The Times*, 'does more than make pictures pretty. It adds information. It tells you about distance and perspective in a landscape; it portrays more vividly than black-and-white'.[34] In the context of the competition with ITV, especially in relation to wildlife television, such characterisation of colour television implied that the representations of nature the BBC offered were more valuable, because more informative, than those of ITV, still broadcasting in black and white. Attenborough's statement encapsulates the notion that the medium is true to nature and that colour representations of wildlife, as seen on BBC2, have more knowledge value than those of the competition. From this perspective, *The World About Us*, as a series of wildlife films in colour, was the BBC's reply to ITV's *Survival*, the all-film wildlife series of commercial television, which kept achieving the best ratings with audiences in the late 1960s. In addition to enrolling animal life in support of specific claims about the social political order, as *Life* exemplified, the case of *TWAU* shows that the spectacle of wildlife could also be put to economic use.

On the whole, both *Life* and *The World About Us* established the cooperation between professional film-makers and scientists through film-making as a way to produce knowledge. Both programmes contributed to turning television watching into a form of participation in the production of natural knowledge, notably through witnessing expert discussion. From this point onward, questions of technical expertise in relation to the handling of the camera, the editing process and post-production in general became central to asserting the cognitive legitimacy of natural history television. This shift is an expression of the BBC NHU's newly

---

[32] Christopher Parsons, Oral history interview, 26 July 2001, Wildscreen.

[33] Richard Wagner, 'Colour TV awaits audience', *The Times*, 2 December 1967, p. 13

[34] David Attenborough, 'American lesson is that ambition pays', *The Times*, 16 November 1967, Colour TV, p. III.

gained self-confidence in its technical ability. Internal discussions during the negotiations that took place with Alan Root in 1966 to secure his contribution to *TWAU* suggest as much. Advising on the arguments Bristol should use, David Attenborough suggested as 'a bargaining point' the 'BBC expertise inextricably involved in the film in the shape of editing, dubbing and recording'.[35] As we consider in the next chapter, this self-confidence shaped the relationship between wildlife broadcasters and life scientists in a way that maintained the latter at the periphery of the film-making process, at best as advisors, or as providers of raw material on which film-makers could exert their expertise. At the end of the 1960s, wildlife broadcasters at the BBC had succeeded in establishing a professional culture of wildlife television making, one informed by science. But this involved delineating a strict boundary, patrolled by technical experts, the natural history film-makers, to maintain the integrity of wildlife broadcasting as a profession.

References

Agar, J. (2008). What happened in the sixties? *The British Journal for the History of Science, 41*(4), 567–600.

Bocking, S. (1997). *Ecologists and environmental politics: A history of contemporary ecology*. New Haven: Yale University Press.

Boon, T. (2015). 'The televising of science is a process of television': Establishing Horizon, 1962–1967. *British Journal for the History of Science, 48*(1), 87–121.

Burkhardt, R. W. (2005). *Patterns of behavior: Konrad Lorenz, Niko Tinbergen, and the founding of ethology*. Chicago and London: The University of Chicago Press.

Latour, B. (1987). *Science in action: How to follow scientists and engineers through society.* Cambridge: Harvard University Press.

MacKenzie, J. M. (1988). *The empire of nature.* Manchester: Manchester University Press.

---

[35]Controller BBC-2, 'Alan Root', memo to Editor Natural History Unit, 17 June 1966, BBCWAC WE21/57/1.

Milam, E. L. (2019). *Creatures of Cain: The hunt for human nature in Cold War America*. Princeton: Princeton University Press.

Morris, D. (1967). *The naked ape: A zoologist's study of the human animal*. London: Jonathan Cape.

Parsons, C. (1982). *True to nature*. Cambridge: Patrick Stephens.

Ritvo, H. (1987). *The animal estate: The English and other creatures in the Victorian age*. Cambridge, MA: Harvard University Press.

Roszak, T. (1969). *The making of a counter culture: Reflexions on the technocratic society and its youthful opposition*. New York: Doubleday.

Sinclair, A. (2012). *Serengeti story: Life and science in the world's greatest wildlife region*. Oxford: Oxford University Press.

Turner, M. (1987). *My Serengeti years: The memoirs of an African games warden*. London: W. W. Norton.

7

From Oxford to Bristol and Back: The Invention of Scientific Wildlife Television

In the 1960s, the epicentre of creativity and innovation in wildlife film-making stood in Oxford. Gerald Thompson, an entomologist, and Niko Tinbergen, a zoologist, both lecturers at the University, had each independently developed an interest in film-making, as much a research tool as a means of communicating scientific research and its results to wider audiences: students and the interested public. Their motives were diverse, ranging from the idealist view that it was scientists' social duty to communicate with non-scientists[1] to more commercial projects. In the latter case, as Tinbergen put it, producing films for the BBC was a means of raising the funds necessary to make as many free copies of educational films as possible available to schools. Niko Tinbergen was eager to make his ethological studies of animals in the wild known outside academic circles. Gerald Thompson, who had specialised in applied entomology, mostly studying insect populations in Ghana in the 1940s, had in the 1960s turned to investigating the use of film as a tool for research. His ambition, then, was

---

[1] Niko Tinbergen, personal letter to Christopher Parsons and John Sparks, 7 September 1968, p. 1. BBCWAC WE8/600/1.

© The Author(s) 2019
J.-B. Gouyon, *BBC Wildlife Documentaries in the Age of Attenborough*,
Palgrave Studies in Science and Popular Culture,
https://doi.org/10.1007/978-3-030-19982-1_7

'to bequeath to posterity a hundred educational films that won't be bettered in fifty years'.[2] In both cases, the main emphasis was on the behaviour of animals, which was also the NHU's stock in trade. But by looking at behaviour from the perspective of the Darwinian theory of evolution, the Oxford film-makers could construct compelling narratives, absent from the more descriptive, natural historical approach favoured thus far in such programmes as *Look*. In addition, taking advantage of their privileged cognitive status as certified scientists, these newcomers in the field of wildlife film-making adopted an approach to filming animals which authorised and valued intervention and the use of controlled conditions insofar as it enabled them to distil the essence of the specific behaviour they wanted to depict on film. With more cognitive clout, unencumbered by the ethos of non-intervention, the cornerstone of amateur naturalists' film-making, these scientists obtained footage the latter were unlikely to get, unless they were very patient and very lucky. Their interventionist approach enabled scientists to produce films faster than such wildlife cameramen as Eric Ashby, at a lower shooting ratio, using less film than was the norm for this type of filming.

Oxford film-makers' transformative influence on the NHU in the late 1960s conjugated with an increased pressure within the BBC to drive production costs down to encourage programme makers to develop a new standard of practices in wildlife television making. The second half of the 1960s witnessed a profound remodelling of the Corporation's organisational architecture, and of the rationale underlying the way it functioned. Until the mid-1960s, the ideology of public service broadcasting drove BBC operations, and broadcasting was conceived of as an instrument of public betterment. After 1967–1968, a logic of corporate management borrowed from the industrial and commercial corporate world took over, sustained by the increased valuing of the notion of professionalism inside and outside the Corporation (Burns 1977). These institutional changes occurred in response to the financial pressure placed on the Corporation by the launch of BBC2 in 1964 and of colour television three years later. To compensate for these large infrastructural investments, the BBC needed

---

[2] Jeffery Boswall, memorandum: 'Contract with Thompson for 3 "looks"', 17 February 1967, p. 2. BBCWAC WE13/1,071/1.

to increase the licence fee but found little support in government circles where it was perceived as 'extravagant'. The 1967–1968 reorganisation along conventional, even traditional, lines in large industrial organisations—notably turning television and radio into 'product divisions'—was partly motivated by the need to diffuse this kind of hostility. But it was also a form of cultural alignment of the Corporation with its competitors at a time when commercial television kept securing larger audiences than the BBC (Burns 1977) was able to do. From this perspective, embracing an approach to wildlife programme making informed by science and authorising a more interventionist approach made sense economically as it enabled reduction of the cost of programme making. The Oxford film-makers provided executives and managers at the BBC with arguments to change the NHU's prevailing approach to film-making. Notably, the work produced at Oxford helped justify moving wildlife television away from amateur natural history toward science.

A close collaboration with the Oxford film-makers forced the NHU into articulating its identity as a centre of expertise for the production of wildlife television programmes destined for non-specialist audiences, as opposed to scientific film-making. To maintain its existence as an entity the Bristol unit asserted its status as an obligatory passage point for scientists willing to address, through broadcasting, non-professional audiences and communicate with films about animal behaviour. Simultaneously, Tinbergen, Thompson and others in Oxford, conceiving of their collaboration with the NHU as a source of income, treated the NHU as a client, unsettling the balance of power which had prevailed between the Bristol unit and its external contributors thus far. With the scientists at Oxford, the balance of power could potentially be reversed, placing the Bristol broadcasters in a subservient position. This forced them to become more assertive of their own expertise in programme making to remain on an equal footing with the scientists. Many of the early films the Oxford academics produced for the NHU examined parasites and their relationship with their hosts, looking at how evolution of the host drove that of the parasite and vice versa. This is also a good metaphor to understand what turned out to be the tumultuous relationship between the Oxford film-makers and the NHU.

Signals for Survival: Breaking New Ground in Wildlife Television

Ethologist Niko Tinbergen, based in the Department of Zoology at Oxford University since 1949, had started, in the late 1950s, to use a motion picture camera for his research on animal behaviour, to the extent that he credited some of his findings to the films he had shot in the field (e.g. Tinbergen 1960: 2; see also Mitman 1999: 71). At the same time as Tinbergen was developing his use of film as a research tool, he had also become adept at using these research films as teaching aids, showing specially edited versions to his students at the end of lectures, 'after I have told the basic story at leisure'.[3] At first, Tinbergen's use of film as a teaching resource was limited to filming patterns of behaviour. However, in 1962–1963, he began using film not only to document his findings but also his research methods, producing what he called research-in-action films. This development coincided with the moment when Tinbergen was conducting research projects in the nature reserve of Ravenglass, in Cumbria, with a group of students (Shaffer 1991; Kruuk 2003). This is also at that time when Tinbergen entered a close working relationship with Hugh Falkus, which would lead to a collaboration with the NHU spanning over a decade.

Hugh Falkus, angler, hunter, naturalist, natural history writer, professional film-maker and occasional contributor with the BBC's NHU, lived in a cottage near Ravenglass. At the time, he was producing a series of short episodes on life in Cumberland for Border Television, a local television station that was part of the ITV network. Having heard of Tinbergen and his students' research at Ravenglass, Falkus paid them a visit. Out of the encounter came four programmes for his series *Five Minutes with Hugh Falkus*. This first contact with television led Tinbergen to see the then still relatively new medium as a good way of making his work known to larger audiences, attracting public attention to the nascent discipline of ethology (Kruuk 2003). Tinbergen was also very taken by Falkus's personality, and together, the two men decided to embark on the production of half-hour films depicting Tinbergen's research and to approach the BBC's NHU

---

[3] Niko Tinbergen, personal letter to Christopher Parsons, 16 August 1968, BBCWAC WE8/600/1.

with them, taking advantage of Falkus's acquaintance with Christopher Parsons. The first two films coming out of this collaboration, *The Gull Watchers* and *The Sign Readers*, were transmitted in June and December 1964, respectively, as episodes in the series *Look*. Both films depicted the research taking place at Ravenglass, on the social life of gulls, featuring some of the experiments Tinbergen and his students conducted, such as introducing tame crows, hedgehogs or stuffed foxes in the gull colony to test the birds' reactions to intruders.

Following the transmission of the second one, *The Sign Readers*, which showed how Tinbergen and his students interpreted animals' tracks and traces, Jeffery Boswall, who that year had taken over the production of *Look* from Christopher Parsons, sent a letter of feedback to Tinbergen. Praising Tinbergen's on-screen performances—'Each of your own performances I and my colleagues enjoyed and admired'—Boswall congratulated him on the very high professional standards of both films 'from the artistic and scientific stand-point', qualifying Falkus and Tinbergen as 'a powerful contribution of whom the audience must see more'. Yet, he also remarked that both films, 'from the technical stand-point, were a fraction below par'. As Boswall explained:

> one or two of your shots were a fraction grainy. No problem should arise from the simple fact that you use colour and not black and white, because colour in our experience translates very acceptably into black and white. But it is vital, as I am sure you already appreciate, to keep the original in absolute pristine condition and to copy directly off that when making a black and white dupe neg. I gather from John Martin that for some reason this wasn't possible and that he had to dupe off a *print* from your Ektachrome original.[4]

In the words of Robert Reid, head of the Science and Features Department at the BBC in 1969, writing in *Nature* on the relationship between television producers and scientists, the 1960s were the days 'in which television constantly bared its breast and its methods in self-analytical dramas, documentaries and comedies' (Reid 1969: 456). To provide Tinbergen and other scientists involved in filming with explanations of the methods and

---

[4] Jeffery Boswall, personal letter to Niko Tinbergen, 12 January 1965, BBCWAC WE8/600/1.

processes of broadcasting was part of the same movement that produced *Unarmed Hunters* (1963; Chapter 5). In this instance, though, these explanations were meant to convey to Tinbergen a notion of the technical constraints associated with television broadcasting, which set television apart from other film-based media, such as educational cinema.

Although quite cajoling, Boswall's letter was meant to impress on Tinbergen the idea that if he wanted to be involved in television broadcasting, the medium had to take precedence over other channels of communication and distribution more habitual to him. This implied adopting elements of the professional culture of television broadcasting. As Robert Reid was proclaiming in his 1969 piece in *Nature*, if a scientist wanted to take responsibility for producing a programme and carry this 'new role well, he has to acquire the professional skill and experience of a producer, and devote a producer's time and energy to his programme. He will cease to be a scientist' (Reid 1969: 458). Tinbergen's main outlet, when he started collaborating with the BBC, was the production of educational films circulated to schools and universities in addition to the films he used in his own classes. As Tinbergen explained to Nicholas Crocker, 'Having only very limited funds available for this, the production of a television film for the BBC is for me a tool, a lever by which the production of such films has become possible.'[5] However, to broadcasters, the visual interpretation of a subject in order to produce a television programme was an activity of its own, different from the production of teaching films. Although to the external person it seemed simple and straightforward, 'no two programmes present the same problems and no simple formula can satisfactorily describe all the processes in which the producer has to be involved' (Reid 1969: 456). To educate Tinbergen in these matters, Boswall relied on Falkus, who had experience in working for television, and to whom Boswall ascribed the task of conveying the broadcasters' 'viewpoint and outlook',[6] which differed from that of the producer of educational films. As these exchanges show, collaborating with Niko Tinbergen provided the NHU with an opportunity to further develop the cultural space of scientifically informed wildlife television, distinguishing

---

[5] Niko Tinbergen, personal letter to Nicholas Crocker, 2 May 1968, p. 2. BBCWAC WE8/600/1.
[6] Jeffery Boswall, personal letter to Niko Tinbergen, 12 January 1965, BBCWAC WE8/600/1.

it, this time, from the kind of educational cinema Tinbergen was involved in.

Following the broadcasting of the first two programmes, Boswall encouraged Tinbergen and Falkus to work on two further films for *Look*. One would eventually evolve into *Signals for Survival*, the award-winning episode of *The World About Us* series—of which more later. The other one, *The Beachcombers*, was broadcast in November 1965. Its central figure was the herring gull, 'the arch-beachcomber', Tinbergen and his students only coming in 'as its colleagues, who look with admiration at his efficiency, … doing some amateurish beachcombing of [their] own'.[7] The gull was contrasted and compared with other animals—fox, badger, hedgehog, curlew—which also made a living from what they found on the beach in Ravenglass. In the *Radio Times* billing that announced the programme, Tinbergen and his students were characterised as 'professional naturalists'. This labelling is part of the NHU's effort to contradistinguish themselves and their programmes from the amateur natural history which had dominated their practice until then (Chapter 5). Tinbergen was already enjoying a degree of notoriety in Britain as the main proponent in the country of Konrad Lorenz's new approach to studying animal behaviour. Labelling him a professional naturalist was a rhetorical strategy to peg the new wildlife television developed in Bristol to ethology, defining it in the public sphere as the professionalization of natural history. It also legitimated wildlife television's almost exclusive focus on animal behaviour at the expense of other practices of natural history, such as collecting and classifying, conversely cast as amateur ones. The relationship between Tinbergen and the NHU was one of mutual exploitation. It helped Tinbergen gain public support for his work and enabled the NHU to renew wildlife television. It lasted for as long as the two parties could each find an interest in it to forward their respective objectives.

This collaboration reached a climax in 1968 with the film *Signals for Survival*. Presenting the social life in a colony of herring gulls, and the way the birds communicate by voice and posture, it was two years in the making. A collaboration between Tinbergen and Falkus, the project began as an all-film *Look* programme but was eventually considered for

---

[7]Niko Tinbergen, personal letter to Jeffery Boswall, 25 November 1964, BBCWAC WE8/600/1.

the new BBC2 series, *The World About Us*. Produced by Christopher Parsons, *The World About Us* had been created by David Attenborough as a
replacement of *Life* (Chapter 6). This series also served to introduce a new
format of wildlife television programme, the uninterrupted fifty-minute
transmission of a film, without studio sequences hosted by a trustworthy personality. Showing colour films, *The World About Us* contributed to
advertising the value of wildlife film-making as a means of exploring and
knowing the natural world, and of linking scientists and non-scientists.
Signals for Survival fitted well in this project. Mostly shot by Niko Tinbergen and Hugh Falkus, the film rested on Tinbergen's research over the
previous twenty years and a close collaboration between the two men:

> Our joint film on Signals for Survival was preceded by first my own twenty
> years' hard work, then by Hugh and me discussing endlessly what it really
> was in my story that was so exciting; then by sketching out a script; then by
> continually adapting the story, and clarifying it, as we saw what we wanted
> and what we could (and could not) hope to show.[8]

To Tinbergen, making *Signals for Survival* was a means of fulfilling what to
him was every scientist's 'urgent social duty' to reach 'the non-professional
public'. Simultaneously, he 'consider[ed] this film just as much a "publication" as an article in a scientific journal',[9] and thus as an intervention
in the scientific field via television. Commenting on his film work for the
BBC Tinbergen stated: 'Don't think that we believe for a moment that we
have done something perfect; but we do think that we are doing something
of a new kind, and on a high level.'[10] For the BBC, to broadcast *Signals
for Survival* was to participate in the scientific conversation, simultaneously enabling a scientist to address his peers and allowing non-specialist
audiences to stand as witnesses to this conversation.

   However, although Tinbergen provided most of the footage and conceived of *Signals for Survival* as his—and Falkus's—film, broadcasters in

---

[8]Niko Tinbergen, personal letter to Christopher Parsons and John Sparks, 7 September 1968, p. 2.
BBCWAC WE8/600/1.

[9]Niko Tinbergen, personal letter to Christopher Parsons, 7 July 1969, BBCWAC WE8/600/1.

[10]Niko Tinbergen, personal letter to Christopher Parsons and John Sparks, 7 September 1968, p. 5.
BBCWAC WE8/600/1.

Bristol saw it largely as the result of the work of its producer, Christopher Parsons. For, whereas Falkus and Tinbergen had shot good footage, to producers at the NHU, 'they're not that fantastic *in TV terms*'.[11] The broadcasters suspected that neither Tinbergen nor Falkus '[were] very aware of other film makers in this field [natural history film-making]', hence their limited understanding of the medium for which they were working. When it came to editing the film and constructing its sound-track (essential for a film presenting the way gulls communicate by voice and posture), Tinbergen was absent from the editing room. In the autumn, Parsons, Falkus and David Aliband, the NHU's film editor, met to work on the final stage of production. Aware that the film's success depended largely on the accuracy of the sound track—'not only for scientific purpose but also in order to create a sense of realism, of actually being *in* [original emphasis] the gull colony'—the three men engaged in what Parsons later described as 'the most careful and detailed pieces of post-synchronisation yet undertaken on a wildlife film at Bristol', recreating the sounds of the gull colony and matching every call and wingbeat to the action in every film shot (Parsons 1982: 262). In July 1969, *Signals for Survival* was the BBC entry for the Italia Prize, one of the most prestigious international awards for television programmes, which it won, in the documentary category. The BBC flew Tinbergen, Falkus and Parsons to Mantua, where the latter received the prize on behalf of the Corporation. David Attenborough, as the controller of BBC2, wrote to congratulate Tinbergen:

> I was delighted when the decision was taken that *Signals for Survival* should be the BBC entry for the Italia Prize, for I truly believe that it is not merely a superb natural history film, different only in degree from many others, but a substantial step forward in the whole genre. ... It has brought the BBC a great deal of prestige, and I must not only congratulate you but also thank you.[12]

---

[11] Richard Brock, handwritten note to Nicholas Crocker, n.d. [May 1968] (original emphasis). BBCWAC WE8/600/1.

[12] David Attenborough, personal letter to Niko Tinbergen, 25 September 1969, BBCWAC T41/434/1.

With *Signals for Survival*, the NHU was claiming new territory, decidedly establishing its new brand of wildlife television, no longer informed by amateur natural history or big game hunting but by science. For his part, Tinbergen saw his film winning the prize as a part of his ongoing effort to advertise ethology as scientifically legitimate: 'Since our science is still fighting (with success so far, but not with as complete success as I wish) for "recognition", I must use the publicity which being entered for the Italia Festival means.'[13] However, just as Tinbergen had been sidelined during the post-production for the film, he was again, when the BBC communicated about getting the award. The press release, reprinted, for instance in *The Times*, announced that the Italia Prize had gone to the BBC for 'a programme on seagulls, directed and narrated by Mr. Hugh Falkus'.[14] Tinbergen took issue with such a communication strategy, writing to Huw Wheldhon, the managing director of BBC Television: 'We were all a little taken aback by the B.B.C.'s own announcement of the Italia Prize. It said everywhere "the B.B.C. announces the awarding of the Italia Prize etc. to *their* film 'Signals for Survival'", or words to that extent, with a remark about seagulls or something'.[15] David Attenborough, tasked with replying, explained that

> the Italia Prize is, in a quite specific sense, a festival in which the participants are not programme makers but broadcasting organisations. The winner, therefore, is never an actor, a director, a writer or a composer, but always a broadcasting organisation.

> …

> Needless to say, when the Award is given, it is, in practice, seen quite clearly to belong to the person or the team who created the prize-winning programme, and we always do our best to make sure that this is apparent by flying the creators out to Italy if we win an Award. In addition, we do our best to make it absolutely clear in our Press Releases where the credit lies. In this particular instance, we stated quite clearly that the programme

---

[13] Niko Tinbergen, personal letter to Christopher Parsons, 7 July 1969, BBCWAC WE8/600/1.
[14] *The Times*, 'Italia prize for BBC TV film', 23 September 1969, p. 6.
[15] Niko Tinbergen, personal letter to Huw Weldon, 25 September 1969, p. 1. BBCWAC T41/434/1.

was introduced and photographed by you, directed and narrated by Hugh Falkus, and presented by Christopher Parsons. The Press, however, in its need to simplify and shorten, simply said that the BBC won the Documentary Prize, and this, though unfortunate because it is so crude a statement, is nonetheless in a technical sense correct.[16]

To carve out the cognitive territory for its own kind of wildlife television, the NHU was eager to appropriate the science but keep the scientists at bay, thus ensuring that wildlife broadcasting visibly remained in the hands of the broadcasters. Scientists could participate but had to do so under the visible leadership of professional film-makers and broadcasters. If they wanted to become broadcasters, they had to cease being scientists and become experts at producing television programmes.

The decade-long collaboration between Tinbergen and the NHU contributed to bringing into existence a renewed approach to wildlife television. With his films offering Darwinian interpretations of animal behaviour, looking at the adaptation value of patterns of behaviour, Tinbergen successfully demonstrated how wildlife film-making could extend beyond the spectacle of the amateur natural history pursuits of collecting, observing and aesthetic enjoyment to embrace a scientifically informed analytical approach to animals. Staging himself and his students at work in his films of 'research in action', he also added a new character to the repertoire available to wildlife film-makers to tell stories about nature: the field scientist. The NHU's collaboration with Tinbergen, a scientist interested in film-making, also helped establish a boundary between the broadcasting institution and scientists, locating the authority over the broadcasting of wildlife television within the broadcasting institution. As the case of Tinbergen shows, scientists eventually were only able to provide the raw material on which broadcasters could exert their expertise in programme making. In this relationship, scientists had to learn to work with broadcasters and to understand the culture of broadcasting, centred on the necessity of addressing non-specialist audiences. But whilst Tinbergen, with his studies of animal behaviour informed by the Darwinian theory of evolution, brought scientific clout and an intellectual grounding

---

[16]David Attenborough, personal letter to Niko Tinbergen, 1 October 1969, pp. 1–2. BBCWAC T41/434/1.

to the NHU's output, another Oxford academic who had developed an interest in filming, Gerald Thompson, provided the NHU with the technical expertise and ingenuity which the Bristol unit was lacking. Thompson enabled the NHU to advance a lab-like approach to filming which complemented the type of scientific wildlife television they were developing at the time.

The NHU Finds the Future of Wildlife Television in Oxford

Bristol's relationship with Gerald Thompson had begun in 1960, when Thompson, together with his assistant, Eric Skinner, had entered and won the BBC's Council for Nature's natural history film competition with their half-hour entry depicting the plight of the alder woodwasp (Chapter 5). The film had provoked a lot of excitement at the Bristol unit. Here was a film-maker who could claim genuine expertise of his topic—insects— and produce original footage of never-before-seen animal behaviour. The winning entry in the 1960 competition, the film was to be broadcast as a *Look* episode. However, the transmission was stalled by a disagreement between Thompson and the NHU about the payment the former should receive. The dispute was foretelling of what would be a major aspect, over the years, of the relationship between the NHU and the film-makers in Oxford, the latter consistently refusing to let the NHU dictate the rules of the relationship, especially financial ones. Eventually, after Thompson managed to extract £150 from the NHU (on top of the £500 prize), having gone as far as writing directly to Kenneth Adam, the controller of the BBC's television programmes, the film was transmitted. And although the NHU could have been expected to blacklist Thompson as a trouble maker, they tried, instead, to put the disagreement behind them and start on a new footing. This shows that the broadcasters in Bristol perceived Thompson's potential contribution to wildlife television as vital to the future development of the Unit.

After the woodwasp film had been broadcast, Eileen Molony, the series producer, wrote to Thompson a conciliatory and most flattering letter,

praising the film and celebrating the way it had been received by audiences: 'The larger section [of viewers] thought the programme evidence
of a fine piece of research combined with superb photographic skill.'[17]
Summarising the Audience Research Report for the *Look* programme in
which the woodwasp film had been shown, Molony asserted, through her
letter, the BBC's role and status as arbiter of television programmes' quality
and value. Thompson, in his reply, although striking a tone of bonhomie,
was keen to emphasise that he did not rely on the BBC to get a sense of
the value of his film effort. Not shying away from stereotypes, he shared
his own anecdotal evidence of good reception by local audiences: 'a group
of trainee nurses at a London hospital ... only interested in the mating
scenes!' or the local postman who 'stopped and remarked "baint you the
two gennelmen who were on Television ...?", apparently he was an appre-
ciative viewer!'. Thompson concluded his letter by revealing that he did
not own a television set, suggesting some lack of interest in, or at least
some distance from, the medium, but that he was 'seriously considering
getting [one]!'. However, building on the good audience reception of the
woodwasp film, Thompson, in the same letter, emphasised his keenness
for further collaboration with the BBC: 'Presumably the audience reaction
was sufficiently encouraging to give hope for future insect programmes?'.
As if trying to whet Molony's appetite, Thompson then described his new
filming project: 'We are trying to complete a short ... film on "Tiger
Beetle". The T.B. is predatory both as grub and beetle; it's a most vora-
cious beast, a veritable tiger of the insect jungle. The result may be too
horrible for presentation to the public!'.[18]

Shortly before Christmas 1961, Thompson eventually paid a visit to
the NHU in Bristol, taking with him some of his footage. His goal was to
create a relationship with the NHU whereby he would supply the Bristol
unit with close-up material and work closely with the Bristol film-makers
as they shot the wide-angle, long shots and other establishing sequences
to go with it. His ultimate aim was to raise money to finance a small
laboratory in which to produce film sequences for a range of commercial
contacts dealing with educational films, visual teaching aids, and so on.

---

[17] Eileen Molony, personal letter to Gerald Thompson, 14 June 1961, BBCWAC WE21/68/1.
[18] Gerald Thompson, personal letter to Eileen Molony, 16 June 1961, BBCWAC WE21/68/1.

From this vantage point, Thompson conceived of his initial collaboration with the NHU as a way of getting funds to invest in new equipment. Collaborating with the Bristol unit was, for Thompson, a means and not an end. If another solution presented itself that could help him sell his film as well, he did not conceal his readiness to change tack. For example, upon hearing that Peter Scott was contemplating developing an insect film studio at Slimbridge, he mused and wondered whether a better solution from a financial point of view would be for him to join forces with Scott and run the film side of the project under the Slimbridge label.[19] Yet, at this early stage, the NHU was his only concrete source of income, and upon returning to Oxford, he left behind a list of twelve programmes, or film suggestions, about insects. Molony passed it on to Christopher Parsons, who then arranged a visit to Thompson in Oxford, 'to talk about the whole question of future television programmes on insect life'.[20] Early in January 1962, Parsons made the trip to Oxford with John Burton, the NHU film librarian, and a keen amateur entomologist.

During the visit, Parsons paid particular attention to the technical details of Skinner's and Thompson's film set-up, which he called their laboratory. To Parsons, Thompson was first and foremost a scientist doing film as part of his scientific practice. What he witnessed convinced Parsons that the kind of specialised filming Thompson and Skinner were engaged in represented the future of wildlife film-making, and the NHU should secure their collaboration without delay. Reporting to his colleagues in Bristol, he insisted on the material means Thompson and Skinner had at their disposal, marvelling at the efficiency it allowed: 'A subject can be brought into the laboratory and filmed in three to four minutes.'[21] To Parsons, the two men had 'tremendous advantages over any other film-maker' trying to do the same kind of work. Their ready access to the metal and wood workshops at the Institute of Forestry, where they could modify at will second-hand equipment to suit their needs, meant that they could

[19] Eileen Molony, memorandum to Head of West Region Programmes, 8 January 1962, BBCWAC WE21/68/1.

[20] Christopher Parsons, personal letter to Gerald Thompson, 21 December 1961, BBCWAC WE21/68/1.

[21] Christopher Parsons, 'Thompson and Skinner—Commonwealth forestry institute', memo to HWRP, 16 January 1962, BBCWAC WE21/68/1.

constantly and quickly adapt their filming methods to overcome specific problems posed by their subjects. And the housing of their filming studio in the same building as their place of work meant that they could devote as much time as they wished to their filming. What he saw in Oxford convinced Parsons that Thompson and Skinner were going to be, in a very short time, 'the leaders' in insect, or macro-cinematography, and that the NHU could not 'afford to let them work for the other channel [ITV]'.[22] Desmond Hawkins, who by then had become the head of west-region programmes, took Parsons's analysis seriously and within weeks, Thompson and Skinner were offered a contract to produce three films for the NHU over the course of three years, to be delivered by the end of 1965.

These films were to be part of Parsons's effort to include complete films in the *Look* series, as Peter Scott's involvement with the programme was progressively phased out. Anxious to check on Thompson's and Skinner's progress, Parsons, in December 1962, went to Oxford again. There he found what to him was, without doubt, one of the 'finer studio equipped for macro-photography in the country'[23] and the two film-makers entertained him with footage he judged to be 'the finest examples of macrophotography' that he had ever seen. Moreover, all the footage Parsons was shown was in colour. In 1962, the question was not whether the BBC would broadcast in colour but when, and wildlife broadcasters had begun stockpiling colour material to prepare for the transition. Finally, in keeping with his original impression, Parsons's visit convinced him that the work for which the two men had been contracted would be finished much sooner than their contract with the BBC stipulated. Judiciously showcasing their technical ingenuity, Thompson and Skinner succeeded in persuading Parsons that their work and expertise would soon be indispensable to the NHU. He accordingly concluded his report with an exhortation: 'I feel that we must hang on to them at all costs.'[24] Many in Bristol agreed with Parsons. At the end of 1963, Patrick Beech, Desmond Hawkins's assistant,

---

[22]Christopher Parsons, 'Thompson and Skinner—Commonwealth forestry institute', memo to HWRP, 16 January 1962, BBCWAC WE21/68/1.

[23]Christopher Parsons, 'Thompson and Skinner', memo to Nicholas Crocker, 10 December 1962, BBCWAC WE21/68/1.

[24]Christopher Parsons, 'Thompson and Skinner', memo to Nicholas Crocker, 10 December 1962, BBCWAC WE21/68/1.

rated their work as being of 'the highest quality and to have considerable value over an extended period'.[25]

The production of their first films had been a learning process for Thompson and Skinner under Parsons's guidance. His regular trips to Oxford, as well as their visits to Bristol to discuss the editing or the recording of the commentary were so many occasions, from Parsons's perspective, to explain to them the processes and requirements of the medium and train them to work for television. For instance, after Thompson had sent the material for the *Look* programme on the tiger beetle, Parsons had to re-cut the film. In a letter to Thompson he detailed:

> What I have done basically is to shift a lot of the seasonal scenes from the beginning to the middle of the film, and also to use some of the sequence of the tiger beetle's world in the second Spring sequence. This has enabled me to break up the close-up sequences with seasonal and habitat shots rather more than was possible with your original order.[26]

But in learning about programme making, Thompson and Skinner became bolder when handling the NHU. By the time a new contract came up for negotiation, they had become more assertive about the fees the NHU should pay them as well as how the rights to their films should be configured in their contracts. This, however, did not go unnoticed, as it became increasingly understood in Bristol that the Oxford film-makers saw the NHU primarily as a source of income and an outlet for publicity. When Thompson wrote to suggest that the creation of a new contract should be postponed, Parsons noted in the margin of the letter before passing it on to Crocker, the Unit's head: 'I feel that if we let the contract go until summer 1965, T. and S. will have more bargaining power. I suspect they are hoping that more money may be available then or that prices will have gone up.'[27] Accordingly, the two men were offered a new contract quickly. And with it, their point of contact with the NHU changed.

[25] Patrick Beech, Colour Fund, memo to A.C. (Planning) Tec., 18 December 1963, BBCWAC WE21/68/1.

[26] Christopher Parsons, personal letter to Gerald Thompson, 27 November 1964, BBCWAC WE21/68/1.

[27] Chris Parsons, note to N. Crocker on letter from G. Thompson to C. Parsons, 14 November 1964, BBCWAC WE21/68/1.

As Christopher Parsons was moving on to new projects, notably a series of programmes with Gerald Durrell, dealing with the Oxford film-makers became the responsibility of another member of staff at the NHU, Jeffery Boswall, who, in 1964, had become the producer for *Look*. Whereas Parsons had approached his role with Thompson and Skinner as a nurturing one, Boswall adopted a more antagonistic attitude, consistently fencing off the territory and putting forward the NHU's expertise. In his correspondence with Thompson, Boswall insisted that television broadcasting was first and foremost about communicating with non-specialist audiences, and repeatedly asserted his and the NHU's property of skill in addressing these audiences. The way this relationship developed indicates that although the NHU producers were eager to marry science with natural history film-making and felt that they had found in Thompson a good source of technical expertise to turn film-making into an actual participation in the sciences, they were also convinced that broadcasting should not become subservient to science and should remain in the hands of broadcasters. From this point of view, broadcasters were experts and broadcasting was their field of expertise. This evolution is in line with the notion discussed earlier that in the first half of the 1960s, NHU broadcasters, like their colleagues across the BBC, were keen on developing as a profession and on being perceived as professionals outside the Corporation. It also shows that NHU's broadcasters' self-fashioning as professionals was happening at the same time as they were constructing a closer relationship with the scientific world. In doing so, they established the separation between scientists and broadcasters, which Robert Reid asserted as natural in his 1969 piece, when he claimed that if a scientist became a television producer, she would 'cease to be a scientist' (Reid 1969: 458). Looking at the relationship between the Oxford scientists-turned-film-makers and the broadcasters in Bristol shows that the latter actively enforced this distinction, which is far from being a given, to fashion and maintain their own social identity.

The exchanges between Boswall and the Oxford film-makers, quite abrasive at times, show that as scientists were developing their capacity for communicating their research to non-specialist audiences, so broadcasters in Bristol were positioning themselves and their institution as necessary mediators between scientists and lay publics. On 20 September 1965,

Boswall paid his first visit to the Oxford film unit. It would be an understatement to say that Boswall and Thompson did not 'click'. Upon his return to Bristol, Boswall penned a rather a bilious handwritten note:

> A tough character! Very very money minded. I suspect a disillusioned academic! Now sees his immortality preserved in his teaching films. Is 48 intends to retire at 60 (not 67) set up studio at home and train 20 year-old son as film-maker.
>
> Really only interested in TV as a source of money. Expects to be 'free of it' when enough money coming in from educational sales. Certainly doesn't 'believe in' TV in any sense.[28]

This belief in the medium, at the core of broadcasters' professional identity, entailed a conception of television as an end, and not simply a means. Not sharing in this belief was to negate broadcasters' identity. In subsequent letters to Thompson, Boswall elaborates on what such a belief entails. Central to it is the notion of popular presentation. As a mass medium, television is meant to reach large audiences. As such its mode of address is necessarily popular, as opposed to a specialised one. Thompson visited the NHU on 20 December 1965. There he had a thorough discussion with Boswall about the requirement of the medium, in Boswall's words 'the "pop" (but not *un*scientific) requirements of a LOOK'.[29] In this meeting, at which both Peter Scott and John Burton were present, the main topic for discussion was a film Thompson was planning on, *The Cabbage and Its Enemies*, examining the various parasites living on the vegetable. Trading ideas with Thompson, in successive letters, about the form the film should take, Boswall further defined the popular approach broadcasters in Bristol were taking to presenting natural history topics on TV. This correspondence enables us to capture, at an early stage of its formation, what would become the dominant culture of wildlife television in subsequent decades.

---

[28]Anonymous, 'Notes on discussions with GH Thompson Oxford 20th September 65', BBCWAC WE21/68/1. Boswall later referred directly to the note indicating that he is the author.
[29]Jeffery Boswall, personal letter to Gerald Thompson, 11 February 1966, BBCWAC WE21/68/1 (emphasis in the original), p. 1.

As a scientist specialising in macro-photography, Thompson valued most the close-up shots he could get with his equipment, making visible what the unaided eye could not see. From his perspective, filming fitted into science as one additional means of revealing the hidden truths of nature. Thompson's performative understanding of the medium implied that a film was primarily a means of demonstrating the cameraman's ability to show things. By contrast, Boswall was concerned with how 'the interest of the general viewer might best be secured'.[30] To broadcasters, film was primarily a mode of address. This difference in understanding entailed a reversal of perspective:

> Going through the notes I took on the day, I see we agreed to drop the emphasis on the Cabbage and its enemies, and to concentrate on the Large White as a potential hero. You had in mind certain coverage of this creature; John [Burton], Peter [Scott] and myself had some additional ideas for the melting pot, arising from audience considerations.[31]

Thompson was entranced with his ability to provide close-up shots of insect life to the extent that he could not conceive of a film as more than a collection of such shots. Yet, as Boswall was explaining to him, an accumulation of these nuggets was in fact undermining their value and interest, even making viewers uncomfortable as they created a feeling of forced proximity with the animal.

> The essential—and unique—strength of your stuff for the series is that fabulous close-up-ness, of course. But it does pose a bit of a problem in claustrophobia after a while. If we treat the story as a chronological one, how often can we pull back wide to relieve the intimacy—and indeed remind people just how close we are getting?[32]

Boswall is offering a lesson in controlled enthusiasm for the technical feat, and in the economy of film, whereby high-value shots, like gems in a jewel, are best valorised when encased in less remarkable sequences. To

---

[30] Boswall to Thompson, 11 February 1966, p. 1. BBCWAC WE21/68/1.
[31] Boswall to Thompson, 11 February 1966, p. 1. BBCWAC WE21/68/1.
[32] Boswall to Thompson, 11 February 1966, p. 2. BBCWAC WE21/68/1.

elicit interest for the story, Boswall suggested drawing on the mundane, on potential viewers' everyday-life experience:

> I still like the idea of the housewife finding a caterpillar in the kitchen, and despite this the husband sorting one out from the cabbage on his plate. Also, we can hardly do a show about this particularly well-known pest without showing one gardener painstakingly removing them, while another relies on dusting with an insecticide.[33]

Finally, Boswall remarked that programmes focused on one species presented the difficulty of finding ways of introducing other organisms: 'One of the difficulties of a single-species LOOK is always how legitimately to work in other organisms.'[34] Using the enemies' approach, and a dose of chummy flattery, Boswall suggested a series of 'rigged' shots:

> How about putting lots of dead Cabbage Whites on a bird table and filming one bird —butterfly in bill—in mid-shot (with table out of shot, of course)? To feed one to a nestling would be easy, presumably. One could even plant one on the edge of, say, a Reed Warbler's nest and press the button immediately after the bird picks it up? John thinks it might be possible to 'rig' a dragonfly eating a Large White, by presenting one to newly-emerged Aeshna. Kid's stuff to T. and S., no doubt.[35]

With the change of interlocutor, the tone of the relation between Oxford and Bristol had evolved. Whereas Parsons had been much willing to secure Thompson's and Skinner's collaboration at all costs, Boswall was much more eager to maintain a clear balance of power. To provide a counterweight to the Oxford film-makers' cognitive authority, stemming from their being professional scientists, Boswall first put forward the NHU's expertise in producing television programmes. However, as a scientist doing specialised filming who knew that his technical expertise was precious to the NHU, Thompson was not ready to submit to Boswall's hectoring. His initial response was to distance himself from the film-making

---

[33] Boswall to Thompson, 11 February 1966, p. 2. BBCWAC WE21/68/1.
[34] Boswall to Thompson, 11 February 1966, p. 2. BBCWAC WE21/68/1.
[35] Boswall to Thompson, 11 February 1966, pp. 2–3. BBCWAC WE21/68/1.

considerations laid out by Boswall, emphasising his knowledge of nature to reassert his control of the relationship:

> I feel that Eric [sic.] and my job this year should be to film the 'guts' of the programme, the life history of the Large White and of Apanteles, that much maligned parasite which is really the hero! ... in the light of what we get this summer we can decide (a) whether to go ahead with the programme (b) if so, how to round it off. I do not wish to become involved in mass rearing of butterflies, for several reasons, not least the horrible smell of the caterpillars! I do not fancy arranged shots unless i) I know for certain that the end product really does take place in nature, ii) the time spent is not exorbitant.[36]

Boswall let the issue subside. Meanwhile, Thompson seemingly lost interest in the prospect of working for *Look* and began planning for an ambitious filming expedition to Jamaica to take place in the summer of 1967. In order to get the expedition funded, he approached Christopher Parsons at the NHU to offer the BBC to contract him and his crew to the tune of £7000, on the agreement that they would film enough material for three fifty-minute colour shows for BBC2.[37] As had become usual with Thompson, the negotiations first stumbled on the questions of rights and ownership of the film material. But in January 1967, Thompson, in need of money, brought up the cabbage butterfly film again, urgently requesting Boswall to come and visit him at Oxford.[38] Boswall travelled to Oxford on 13 February 1967. What he saw there did not quite meet his expectations in terms of progress, as Thompson could only show him his trademark macro-cinematography footage. Despite the quality of the material—'It is of compelling intimacy, and high interest value so far as it goes. No-one else we know can produce stuff like it'[39]—Boswall remained unsatisfied with it as far as the grammar of television went:

[36] Gerald Thompson, personal letter to Jeffery Boswall, 15 February 1966, BBCWAC WE21/68/1.

[37] See, for example, Christopher Parsons, 'Thompson and Skinner: Jamaica expedition', memo to Miss Mimi Cooper, TV Enterprise London, 26 August 1966. BBCWAC WE21/68/1.

[38] Gerald Thompson, personal letter to Jeffery Boswall, 17 January 1967, BBCWAC WE13/1,071/1.

[39] Jeffery Boswall, personal letter to Gerald Thompson, 17 February 1967, p. 3. BBCWAC WE13/1,071/1.

It does, however, by its nature, and by the fact that you tend to suggest somewhat academic animals, bring with it certain problems in popular presentation. The need for wide angles, to remind people of just how privileged a view they are getting, the need for a faster pace than may be necessary for other purposes, the need to relate to the ordinary person's experience, etc., etc.[40]

In the meeting, Thompson questioned Boswall's expertise and authority as a producer, and so the latter felt compelled to assert both, exemplifying the central role of the producer in the professional culture of wildlife television as it was developing at the NHU in the 1960s:

> I am *not* unsympathetic to the viewpoint of the cameraman-director, but I am employed as a producer. And if material offered is not of suitable content, in my judgement, then I must say so. If I'm consistently wrong, I'll get the sack, and rightly. But in the case of the Cabbage White stuff filmed so far, superb though it is in certain ways, it could not of itself by any stretch of standards or generosity be the exclusive basis for a 25-minute LOOK.[41]

This exchange shows that as they were developing a relationship with another professional body—scientists—wildlife broadcasters defined their own professional standards and their identity as encompassing both television production and natural history skills. In his exchanges with Thompson, not only did Boswall continuously emphasise the exigencies of broadcasting but he also provided Thompson with repeated evidence of his mastery of scientific knowledge, naming animals using Latin binomials, or highlighting the NHU's connections with the networks of science—for example, offering suggestions on how Thompson could represent on film the migration of the large white, Boswall noted:

> I hope very much that you will be willing to take up the challenge of this migration business. John [Burton] who has written papers on insect

---

[40]Jeffery Boswall, personal letter to Gerald Thompson, 17 February 1967, p. 3. BBCWAC WE13/1,071/1.
[41]Jeffery Boswall, personal letter to Gerald Thompson, 17 February 1967, p. 3. BBCWAC WE13/1,071/1.

migration, and Robin Baker, of the Zoology Department at Bristol, who is spending three years on the migration of *Pieris brassicae*, would both be very willing to advise on how the thing could be most economically achieved.[42]

Addressing Thompson as a cameraman rather than a scientist, Boswall presents the NHU as a node in a network from which the cameraman can get scientific advice. Reversing the relationship of cognitive authority, Boswall signals that if the NHU can sort out the science, scientists can't sort out addressing audiences.

These contests of authority did nothing to improve an already thorny relationship. In a long memo summarising the state of the situation with Thompson after two years, Boswall shared his view that the relationship was worsening, as he could not get the scientist to share the broadcaster's standpoint. Thompson, he explained to Nicholas Crocker, 'is *not* interested in the thing we are *exclusively* interested in: popular television presentation'.[43] Instead, Thompson saw his relationship with the NHU as him shooting educational films which he hoped the NHU would find adaptable to television, rather than shooting a film to the NHU's requirements that he could subsequently adapt for schools. To Boswall it was evident that Thompson saw the NHU as a means and not an end. The conversations between Thompson and Boswall highlighted the incommensurability of their understanding of film. Where Thompson saw two minutes' worth of film, Boswall saw only forty-five seconds: 'He seriously suggested this week that the hatching of Cabbage White pupae from their shells which they then eat was worth all of the 225 ft. (2'24") he showed me. It is worth 45 seconds at most'.[44] In this memo as in previous correspondence, the broadcaster establishes a strong boundary between scientists and broadcasters, suggesting that their interests diverge, and that the latter's expertise, based on a belief in the necessity to address popular audiences, takes precedence over that of scientists.

---

[42]Jeffery Boswall, personal letter to Gerald Thompson, 17 February 1967, p. 3. BBCWAC WE13/1,071/1.

[43]Jeffery Boswall, 'Contract with Thompson for 3 Looks', memo to Editor NHU, 17 February 1967, p. 1. BBCWAC WE13/1,071/1 (original emphasis).

[44]Jeffery Boswall, 'Contract with Thompson for 3 Looks', memo to Editor NHU, 17 February 1967, p. 1. BBCWAC WE13/1,071/1 (original emphasis).

Conclusion

In the 1960s, the NHU found in two Oxford scientists who'd taken up filming, Niko Tinbergen and Gerald Thompson, key allies whose approach renewed wildlife film-making in Britain and enabled the NHU to further develop its own brand of scientific wildlife television. Tinbergen contributed intellectual foundations with his films intended to popularise the new science of ethology, based on the concept that patterns of behaviour had an evolutionary significance. Gerald Thompson, with his assistant, Eric Skinner, developed a brand of technically specialised film-making that publicly transformed the camera into a laboratory instrument, and the television screen into a window on the laboratory. To broadcast Tinbergen's and Thompson's film work enabled the NHU to cast more light on their association with the scientific world and raise the cognitive profile of wildlife television. The advertising coming out of Bristol for the programmes based on those scientists' film work shows a keenness to use scientists' involvement to present the film-making apparatus, in which television was included, as a means of exploring the natural world and producing new knowledge about it. For example, the billing in the *Radio Times* for the tiger beetle film, shown in a *Look* episode in June 1965, emphasised Thompson's camera's positive contribution in renewing spectators' visual perception of nature: 'Natural history film-makers are opening up a new world—a world of strange, fascinating, and sometimes terrifying creatures. Seen through the naked eye they are tiny and harmless, but in close-up they are monsters.'[45] However, in this piece, Thompson and Skinner are presented as film-makers working in a scientific research institute, rather than as scientists doing camera work. To broadcasters at the NHU it was important to enforce the notion that one could not be a scientist and a broadcaster at the same time, and that only the latter could take charge of broadcasting and offer the kind of spectacle television viewers expected from the medium.

Tinbergen and Thompson did not merely influence the development of wildlife film-making at the NHU, though. Through their use of film as a research and teaching tool, they also fostered a culture of film-making

---

[45] *Radio Times*, 10 June 1965, p. 23.

as part of scientific practice in the academic milieu in which they evolved: the Zoology Department at Oxford University. As such, both were instrumental in the establishment of a film unit specialising in biological filming there, Oxford Scientific Films (OSF), Tinbergen and Hugh Falkus as associate members,[46] Gerald Thompson as a founding one. The NHU, because it enabled Thompson first and then the other founders of OSF, Peter Parks, John Paling and Sean Morris, to establish their credentials as specialised film-makers, was equally instrumental in the foundation of OSF. In return, these scientists-turned-full-time-film-makers, through their collaboration with the NHU, contributed to solidifying the shift in Bristol from an approach to wildlife television informed by natural history to one informed by science. The 1972 *Horizon* film *The Making of a Natural History Film* captured this symbiotic relationship, which is explored in the next chapter.

References

Burns, T. (1977). *The BBC: Public institution and private world*. London: Macmillan.

Kruuk, H. (2003). *Niko's nature: The life of Niko Tinbergen and his science of animal behaviour*. Oxford: Oxford University Press.

Mitman, G. (1999). *Reel nature: America's romance with wildlife on film*. Cambridge, MA: Harvard University Press.

Parsons, C. (1982). *True to nature*. Cambridge: Patrick Stephens.

Reid, R. W. (1969). Television producer and scientist. *Nature, 223,* 455–458.

Shaffer, L. (1991). The Tinbergen legacy in photography and film. In M. S. Dawkins, T. Halliday, & R. Dawkins (Eds.), *The Tinbergen legacy* (pp. 129–138). London: Chapman & Hall.

Tinbergen, N. (1960). Comparative studies of the behaviour of gulls (Laridae): A Progress Report1. *Behaviour, 15*(1–2), 1–69.

---

[46]Niko Tinbergen, personal letter to Christopher Parsons, 7 July 1969, BBCWAC WE8/600/1.

8

Oxford Scientific Films: From Field Craft to Film Craft

The films of Niko Tinbergen and Gerald Thompson were a prelude to the development of Oxford University as a centre of innovation in wildlife film-making. From 1969 onward the NHU's main interlocutor in Oxford was a specialised film unit going by the name of Oxford Scientific Films (OSF). Although OSF owes its existence to Tinbergen's and Thompson's influence in the Department of Zoology at Oxford University, it could not have been founded had it not been for the BBC regularly commissioning work from it, even before it had been formally established. OSF cameramen moved away from the hands-off approach to filming which such amateur naturalist cameramen as Eric Ashby favoured. They developed, on the contrary, very hands-on methods, specialising in filming reputedly difficult subjects from unusual angles and always under controlled conditions. To field craft—the knowledge and skills to behave in the field required from any aspiring amateur naturalist film-makers—OSF film-makers substituted what could be called film craft—the skills and ingenuity required to design imaginative apparatus and filming techniques.

Throughout the 1970s, OSF came to define the standard of scientifically informed wildlife film-making. As Paul Crowson noted in his detailed

© The Author(s) 2019
J.-B. Gouyon, *BBC Wildlife Documentaries in the Age of Attenborough,*
Palgrave Studies in Science and Popular Culture,
https://doi.org/10.1007/978-3-030-19982-1_8

study of the origins and establishment of OSF, the BBC 'regarded OSF as important to the industry because their intellectual example was pushing all other nature film-makers in the same direction' (Crowson 1981: 70–71). OSF demonstrated the validity of the belief that film-making could be a way of conducting biological research by other means, simultaneously sharing findings with larger, non-specialist audiences, and creating a television version of biologists' 'expert mode of viewing' (Curtis 2015) which could then be turned into entertainment. As far as the NHU is concerned, the development of OSF first prompted producers in Bristol to reflect on what differentiated them, as broadcasters, from OSF, a specialised film unit originating in the world of academia. This self-examination led to a contest of expertise between Oxford and Bristol. At the same time, under the influence of OSF, Bristol adopted some of their mindset, most notably their belief in the performative dimension of wildlife film-making, understood as being about creating favourable material conditions for nature to perform itself on-screen. Key to this process of acculturation was Mick Rhodes, the head of the NHU, whom David Attenborough had appointed in 1972. This chapter is largely devoted to looking at OSF, how it came about, its relationship to the NHU, and its contribution to the history of wildlife television in Britain.

Although OSF was formally established in 1968, its founding members started operating earlier. OSF's origin story usually situates the actual beginning of the unit in 1966 (e.g. Crowson 1981), when Peter Parks, John Paling and Sean Morris, all three from the Zoology Department of Oxford University and future founding members of OSF, began filming the different phases of the life cycle of the rabbit flea for an episode of *Life*, using special filming equipment and techniques of their own design. The programme, produced by Barry Paine, was to feature Miriam Rothschild's research on the insect's life-cycle and behaviour. In 1966, Rothschild, a noted biologist, had just demonstrated that a rabbit's life-cycle and that of the fleas it carries were synchronized. Namely, when rabbits reproduce, hormones and temperature changes in their bloodstream prompt fleas to breed and reproduce. The host's hormones controlled the reproduction cycle of its parasite. As a doe gives birth to a new litter, new hormone changes prompt the fleas living on the rabbit to migrate onto its nose so that when it licks the new-borns, the fleas can jump onto them. There they mate and lay eggs. The story was quite compelling. Rothschild was a good studio performer, yet her research was new, and no film sequences existed that could be used to illustrate the programme. Filming such minute animals required specific skills and equipment. Peter Parks and his colleagues, who at the time were developing both, are often presented in this version of the story as providential characters, solving a problem for the NHU (Crowson 1981). Yet, in 1966, Peter Parks had been in the orbit of the NHU for a few years already. There, he relentlessly canvassed producers, who kept commissioning work from him. Another possible version of this story therefore could be that producers at the NHU did the flea programme because they already knew of Peter Parks' work and were keen on using it on television.

Beginning as a Caption Artist

In January 1962, Bruce Campbell, head of the NHU in Bristol, received 'a remarkable selection of drawings' that included 'enormously magnified drawings of insects'.[1] The sender was Peter D. Parks, an undergraduate at Keble College, Oxford University. With the drawings came a long letter in which Parks first expressed his appreciation of the work conducted at the NHU: 'I am a great admirer of your many Natural History and Biological programmes, & in fact, all your scientific films. They are beautifully prepared, extremely well put over & include some most intriguing features & aspects of natural science.'[2] Yet Parks could identify a gap in the already large array of topics the NHU covered in its output: 'It occurred to me that there was possibly a niche to be filled, by something about some of the lesser forms of life.' He then went on to effectively describe what would become the speciality, half a decade later, of OSF: 'The sort of thing of which I am especially thinking is in part microscopical, but does not involve very high power magnification.' The drawings accompanying the letter illustrated what Parks meant. He then suggested how his technique could be used for filming.

---

[1] Senior Producer, Natural History Unit, Bristol, 'P.D. Parks', memo to Miss F. R. Elwell, 13 February 1962, BBCWAC WE13/958/1.

[2] Peter Parks, personal letter to Bruce Campbell, 29 January 1962, BBCWAC WE13/958/1.

If it were thought to be of value to show such material alive, even that could be done, I think. Such organisms are easily projected by means of a microprojector onto a screen, from where possibly further filming could be done.

This letter conveys a sense of Parks's technical ingenuity, what would become the driving force propelling OSF. It is also indicative of the professional atmosphere in which Parks was evolving at the Department of Zoology of Oxford University, which employed him as a part-time biological illustrator. There, the idea that innovative techniques of visual representation should be integral to research and teaching was mainstream. Alister Hardy, head of the department, was a leading proponent of what he called new natural history, quantitative, experimental, as opposed to the old, descriptive one. Hardy, who advocated a close association between field work and the lab, had brought Tinbergen to Oxford, as part of his efforts to place the study of living animals in the field on an equal footing of biological prestige as physiology (Burkhardt 2005: 330). Shortly after Tinbergen's arrival in Oxford, Hardy helped him purchase the Bell & Howell HR-1 camera which Tinbergen used 'to bring the excitement of methods and results of the field camp work back to the unfortunates who had been left behind in the lab' (Shaffer 1991: 134). Thompson was another academic embracing this approach, devising inventive techniques and apparatuses in his filming-studio-cum-laboratory to reveal hitherto hidden aspects of nature and investigate the behaviour and life-cycle of life forms too small to be seen with the unaided eye. Both Tinbergen and Thompson firmly tied film-making to the scientific exploration of the natural world, as a tool for research and a means to convey findings to non-specialist audiences. In addition, Tinbergen's and Thompson's film-making activities made it usual for the Zoology Department to dedicate part of its budget to film-making equipment and film projects. Tinbergen's *Signals for Survival*, just like some of the equipment Thompson had used to shoot the *Alder Woodwasp* and subsequent films had been partly funded with money from the Department of Zoology.[3] No firm boundary existed in this academic community between film- and science making.

---

[3] Niko Tinbergen, personal letter to Nicholas Crocker, 2 May 1968, p. 2. BBCWAC WE8/600/1. Crowson (1981: 21).

Tinbergen's and Thompson's successful ventures into wildlife television established film-making as a legitimate career choice for science graduates. Parks's identification of a niche to be filled in the NHU's output indicates a willingness to emulate them.

Upon receiving Parks's letter in 1962, Bruce Campbell did not acknowledge his suggestions for new areas of wildlife television programming. But impressed by the drawings, he forwarded the set to Felicia Elwell, a senior producer in charge of natural history programmes at BBC Television Schools Service. Campbell suggested that although Parks had 'not drawn anything for television yet, … he might have great potential as a caption artist'.[4] And so for the next four years, Parks regularly contributed images of magnified minute life forms in natural history television programmes for the BBC Television Schools Service. He also provided the illustrations for the printed publications which often accompanied these programmes. Through this regular collaboration, he developed a taste for the rapid turnover of television: 'I enjoy the work and the immediate use to which it is put is rewarding.'[5] Then, in January 1966, the NHU, in the person of Christopher Parsons, got back in touch with him.

Parsons was introduced to Parks during a visit to Gerald Thompson in early January 1966 to discuss plans for the use of the latter's footage of spiders for a fifty-minute film, *Walk into the Parlour* (1966). Broadcast on 23 April 1966, as part of the BBC's special programmes for the second edition of *National Nature Week*, it looked at the hunting and prey-capture techniques, courtship and mating behaviours of several species of spiders. Thompson, however, had not been able to film the sequence dealing with how spiders mate. He introduced Parks to Parsons with the idea that some of Parks's drawings could replace the missing footage. Back in Bristol, Parsons commissioned Parks for three drawings, of 'spiders nearing completion of the tiny sperm web', of 'a spider shaking or rubbing [its] opisthosoma on the web and dropping the semen', and of a 'spider dipping one of its pedipalpi into the drop of semen'.[6] In his letter Parsons was eager to establish from the start that Parks should see him as a cognitive

---

[4]Bruce Campbell, 'P.D. Parks', memo to Miss F. R. Elwell, 13 February 1962, BBCWAC WE13/958/1.

[5]Peter Parks to Richard Brock, personal letter, 15 April 1966, BBCWAC WE13/958/1.

[6]Christopher Parsons, personal letter to Peter Parks, 14 January 1966, BBCWAC WE13/958/1.

equal. Not only did he demonstrate his command of the technical terms used to designate the anatomical parts of the spiders; he also asserted his knowledge of the research literature: 'However, I realise that there is little literature, and, as far as I can make out, no visual material exists at all, apart from the odd diagram.' To this familiarity with the scientific aspect of the programme, Parsons added his film-making expertise. The commission included formal instructions about the size and the composition of each drawing, based on how he planned to use them in the film. For example, in the first drawing, spiders

> should both be near the centre of the frame and quite small in order that I can start wide on the general habitat scene and move the camera slowly in. Assume end frame 3″x 4″ with spider and web occupying about one-third of frame, and complete picture about 12″x 9″.[7]

In 1966, the Bristol broadcasters had learned their lesson from five years of a trying relationship with Gerald Thompson and were keen to lay proper foundations for their collaboration with Parks. The letter, combining instructions on the content and the form the drawings should take, strove to establish a relation in which Oxford academics were expected to acknowledge both the scientific literacy of their interlocutors in Bristol and their technical expertise at producing films destined for large non-specialist audiences.

Parks was keen to demonstrate what his expert contribution could be. A diligent worker, he completed the drawings in less than two weeks. In his cover letter, he struck a conciliatory tone, signalling his awareness of the practical constraints attached to television production: '[I] have made close up still not unlike Gerald's close ups of this species so as to help with continuity.'[8] But, when explaining how he'd gotten around the dearth of information about the precise species of spider he was supposed to represent, *Lycosa nigriceps*, Parks also put forward his certified scientific expertise. Given the absence of established reference, his drawings were 'a good approximation of the facts' for 'a generalised Lycosid-type

---

[7] Christopher Parsons, personal letter to Peter Parks, 14 January 1966, BBCWAC WE13/958/1.
[8] Pater Parks, personal letter to Christopher Parsons, 27 January 1966, BBCWAC WE13/958/1.

spider', while avoiding committing 'to uncertain facts'.[9] His degree in Zoology from Oxford University granted Parks a level of interpretative and creative freedom when representing natural facts, a detachment from certainty which would not have been available to an amateur. His scientific certification provided Parks with an expertise in dealing with uncertainty, as opposed to Parsons who had been keen to provide evidence of his factual knowledge. In subsequent exchanges, this contest of expertise continued with Parsons making sure that the balance of power in the relationship between Bristol and Oxford remained tipped toward Bristol. But in the meantime, Parks had begun developing an interest in what he called his 'high-power filming of small & very small animals'.[10]

Filming Microscopic Life Forms

It was after jointly attending a seminar at Oxford where Gerald Thompson had screened his tiger beetle film that Peter Parks began discussing with John Paling the possibility of using film as a means of investigating aquatic micro-life, the topic they were both interested in (Crowson 1981). By then, Parks had spent a few months taking apart a Victorian Ross microscope and building an 'optical bench', a contraption linking together a camera, a magnifying device, and a source of light, with an observation platform. Placing the subject to be filmed on the same rigid framework as the camera, the vibrations occasioned by the camera's motor were transmitted to the subject, thereby maintaining its position relative to the camera. The optical bench thus ensured the stability of the image, even at high levels of magnification (Thompson et al. 1981). The Ross microscope Parks had dismantled to assemble his optical bench included a dark-field condenser. This part enabled him to film using dark field illumination, a lighting technique much favoured by Victorian microscopists to reveal the internal structures of translucent aquatic microorganisms. Indeed, the condenser concentrates the light on the organism but then

---

[9] Peter Parks, personal letters to Christopher Parsons, 18, 21, and 27 January respectively, BBCWAC WE13/958/1.

[10] Peter Parks, personal letter to Ronald Webster, 16 April 1966, BBCWAC WE13/958/1.

diffracts it so that light does not reach directly into the lens. The only light reaching the lens comes through the organism itself, which thus appears translucent against a dark background. Adapting this technique of microscopic illumination to filming was one of Park's and Paling's main innovations.

Using Parks's optical bench, the two men began experimenting with filming aquatic micro-life forms, setting out from the start to produce 'something saleable' (Crowson 1981: 11), and soon setting up a company, Paling-Parks Productions. By mid-1966, they had produced what they called their show reel, which they used as an advertisement for their efforts. This collection of close-up shots of various parasites, mites, and lice, depicted the parasites' behaviour as well as, thanks to the dark background illumination, their internal structures. For example, in shots of *Argulus*, a fish louse, filmed attached to a stickleback's tail, they managed to capture the blood flowing in the blood vessels in the fish's tail and passing into the louse's digestive system. The reel also contained wider establishing shots enabling viewers to appreciate the parasites' relations with their host, their relative sizes and the degree of magnification employed in the film. With this reel, Parks and Paling demonstrated to potential clients their technical ability, and that they shared in NHU producers' understanding of film and its grammar when used on television. Recall Jeffery Boswall's hectoring Gerald Thompson about the necessity of treating close-ups obtained through technical virtuosity as nuggets. Extraordinary shots, like gems, need to be set and interspersed with less striking ones to enhance viewers' perception of their actual value.

With his film work well underway, Parks got in touch with Richard Brock, who was then an assistant producer on the *Life* series: 'In three month time my partner in crime (JOHN PALING) & I will have more or less completed, three ½ hour films on animals—mostly small subject (very small) & mostly never before ever filmed—let alone in colour.'[11] To further emphasise the exceptionality of his footage, Parks insisted that 'much of the animal life we film is aquatic & this perhaps makes it a bit outside the run of normal things'. Eventually, Brocks offered to meet the

---

[11] Peter Parks, personal letter to Richard Brock, 15 April 1966, BBCWAC WE13/958/1 (original emphasis).

two aspiring film-makers in Bristol. They brought with them their show reel, which so much enthused Richard Brock that he strongly encouraged his colleague Barry Paine to make the trip to Oxford and meet with the duo. In Oxford, Paine was likewise treated to a projection of the show reel. Trained as a marine biologist, he still remembered many years later how enthralled he had been by the footage of *Argulus*, shot in dark field illumination and showing the blood passing from the host to its parasite:

> I remember seeing a fish-louse on the tail of a stickleback which John Paling showed me, he was a fish parasitologist and I was just knocked out by this, I thought it was absolutely wonderful. As a marine biologist I had always wanted to photograph plankton.[12]

In autumn 1966, rumours about the film work conducted in Oxford started circulating through in the NHU in Bristol, and soon, many there contemplated working with Parks and Paling. The first to do so was Barry Paine, who in September 1966 had begun planning the episode from the *Life* series about the research conducted by Miriam Rothschild on rabbit fleas. He first took Parks, Paling and their optical bench to Miriam Rothschild's farmhouse at Oundle, where they spent a week in October filming flea-infested rabbits. They had to slightly modify their apparatus, which thus far was designed to film through a microscope, to accommodate a live rabbit on their filming platform, and be able to move the camera around the animal to get views from different angles, all the while preserving the possibility to use dark field illumination on the fleas. During their stay at Oundle, they managed to get full-frame footage of fleas feeding on adult rabbits and mating on new-born rabbits, as well as the spectacular migration of the fleas from adults to youngsters just after their birth. Over a few weeks, first at Oundle and then in their film studio, the Oxford film-makers thus obtained a complete visual account of Rothschild's flea story, producing footage of never-before-seen animal behaviour which could then be shared with a wide audience of non-specialists as well as with the researcher herself (Thompson et al. 1981).

---

[12]Barry Paine, Oral history interview, 31 January 2001, Wildscreen.

When, in the first week of December, came the time to record the flea programme in a studio in Bristol, Peter Parks presented his film as well as the equipment used to obtain it. For a few minutes, he went into some details about the optical bench, the necessity to suppress vibrations when filming at high magnification, and the technique of dark field illumination. This sequence invited viewers to marvel first at the pictures and then at the cameraman's ingenuity. Next, Desmond Morris prompted Miriam Rothschild to provide a running commentary of the film as a way of explaining her research and findings to viewers. In this instance, Morris had Rothschild state that seeing the film had helped her make progress in her work, thereby simultaneously positioning the film work conducted at Oxford in two complementary domains of scientific practice: research and its communication, with the BBC's NHU standing at the junction.

The flea programme established a distinction between OSF's filming and standard scientific film-making. To participate in the programme, Barry Paine had commissioned Eric Lucey from the Research Film Unit at the institute of Animal Genetics of the University of Edinburgh to provide him with high-speed cinematography recordings of fleas jumping. At Edinburgh, Lucey's task was to develop filming techniques to assist researchers studying biological processes best visualised when slowed down or speeded up, such as cell division. Lucey's work thus belonged in the tradition that conceived of film as a scientific tool to investigate phenomena of which movement is an essential aspect, with no interest in the communication of research results to non-specialists. This tradition has its roots in the foundational work of Etienne-Jules Marey and Jules Janssen, who both developed photographic devices with the purpose of capturing movement (Canales 2011; Dagognet 1992). On the set of *Life*, Eric Lucey explained his filming method, insisting on how 'unnatural' it was. His main tool was a Fastax camera equipped with a revolving prism shutter, capable of taking exposures at the rate of up to 8000 frames/sec, a camera which was rare in the UK in 1966. The main technical difficulty in capturing a flea's jump on film comes from the brevity and the unpredictability of the phenomenon. When filming at high speed, a large quantity of film is exposed every second, which makes it very expensive to let the camera run for a length of time in the hope that a jump will occur. Adopting a probabilistic approach, Lucey placed several fleas in a square glass cell of 6 mm in width, assuming that at any time at least one flea would be jumping. Three seconds of filming, using approximately 100 meters of film, recorded twenty jumps.[13] In order to be able to manipulate the fleas, Lucey had first placed them on ice, which had the effect of stunning them. But the heat coming from the intense lighting necessary to film subsequently stimulated them. Barry Paine had been to Edinburgh to film Lucey whilst he obtained his own footage. This making-of sequence showed the insects resting on a bed of ice and then being lifted with tweezers to be placed in the filming cell. Both the setting—a glass cell—and the method employed to obtain the footage—which involved freezing the fleas—disqualified Lucey's film to be defined as a natural history film, for it rested on overt manipulation of the animals and showed them in a context far removed from their natural environment.

Despite being untrue to nature, Lucey's film derived cognitive authority from its context of production, a research institute, and because it served as a basis for the work of a scientist, Henry Bennet-Clark, who, using the film, had managed to build a mechanical model of a flea to analyse the mechanics of the insect's jump. Like Lucey's footage, Parks's had enabled a researcher, Miriam Rothschild, to visualise her object of study and obtain more knowledge about it. But by contrast with Lucey, on the set of *Life*, Parks recounted the time and effort it had taken him and his colleagues to film the flea behaving naturally in its normal habitat, a live rabbit. By emphasising the patience necessary to obtain the images shown to television viewers, Parks summoned the cardinal virtue of natural history film-making, which successive wildlife film-makers—from Cherry Kearton to Eric Ashby and beyond—extolled throughout the twentieth century. To demonstrate patience in filming was to provide evidence that the phenomenon being filmed was a naturally occurring one, and that the film-makers did not intervene in it. Although the filming methods Parks and his colleagues used were very interventionist—placing rabbits on observation platforms under intense lighting—the contrast with Lucey's film characterised Parks's intervention as non-problematic because it did

[13]H. C. Bennet-Clark and E. C. A. Lucey, 1967, 'The jump of the flea: A study of the energetics and a model of the mechanism', *Journal of Experimental Biology, 47,* 59–76.

not detach the flea from its natural habitat and could therefore claim to be truthful to nature.

The rabbit flea programme was a watershed for the development of wildlife film-making in Oxford. Writing to Parks and his colleagues after the programme had been transmitted, Ronald Webster expressed how delighted he was 'to have established the quality of [Parks's] film work.'[14] Indeed, the quality of the contribution had not escaped the attention of the producers at Bristol who soon were sharing 'glowing reports'[15] about the film work conducted at Oxford, many wanting to work with them. For instance, Jeffery Boswall enquired whether Parks would consider providing him with material for his *Look* series:

> This is just to say that having had such glowing reports of the originality and excellence of your film from Barry Paine and Chris Parsons also, I would be very happy to investigate with you the possibility of a contract.[16]

Very soon after the flea programme, though, Parks, Paling, and Sean Morris—who were now trading as Oxford Biological Films—were recruited by Gerald Thompson for a three-month 1967 filming expedition to Jamaica, the prelude to the formation of OSF.

Filming Plankton in Jamaica: 'The Biggest Breakthrough in Nature Cine Photography We Have Had for Some Time'

Gerald Thompson had begun, in late spring 1966, to float the idea of this large-scale filming expedition to the West Indies during conversations with Christopher Parsons. In August that year he formally approached the NHU, offering to go and shoot enough material for three fifty-minute colour programmes for BBC2 the following summer. Thompson, well

---

[14]Ronald Webster, personal letter to Peter Parks, 13 December 1966, BBCWAC WE13/958/1.
[15]Jeffery Boswall, personal letter to Peter Parks, 8 December 1966, BBCWAC WE13/958/1.
[16]Jeffery Boswall, personal letter to Peter Parks, 8 December 1966, BBCWAC WE13/958/1.

informed and increasingly BBC savvy, had timed his project to coincide with the scheduled gradual introduction of colour broadcasting on BBC2. But with this expedition, his ambition went beyond merely producing three colour television programmes for the BBC. As he explained to Christopher Parsons during his year-long bargain with the BBC over who should keep the master of the expedition film and how the rights for the film should be apportioned,

> The end product of my filming, or shall I say the product which justifies my work in this field, is the 16 mm educational film and the 8 mm film loop. ... [M]y object is to produce teaching films of high standard. The alternative approach that I run the expedition purely for the production of TV programmes is also unattractive since the three graduate zoologists who are accompanying me would not then be interested.[17]

Less forthcoming was also the idea that, in addition to gathering enough material to produce educational films, the expedition would help lay the foundation of a specialised biological film unit which would be financially independent from Oxford University. To this end, Thompson wanted to try out a full-scale collaboration between his team—his son David, Skinner and himself—and the three graduate zoologists—Parks, Paling and Morris—already operating through Oxford Biological Films, the film unit placed under the aegis of the Department of Zoology. Ultimately, the expedition's outcome was meant to demonstrate the proposed production company's capability to potential funders. The Jamaica expedition, as it came to be known, was a way of having the BBC cover the production costs of this showcase material.

During the stay in Jamaica, Thompson regularly sent batches of footage to Parsons, who in return would send detailed viewing notes and comments, in effect providing Thompson with professional advice on the quality of his footage and directing the cameramen's work from Bristol. For instance, birds, Parsons wrote,

> were grossly over-shot ... and I think you have wasted a good deal of footage in this way. Once you have got a shot—say, a MLS [Medium Long Shot]

---

[17] Gerald Thompson, personal letter to Christopher Parsons, 8 August 1966, BBCWAC WE21/68/1.

of a certain species of bird flying into a particular perch two or three times and you are reasonably sure that it is in focus, there is no point in going on for another three or four times.[18]

Looking at the rushes as they came in, Parsons was trying to figure out storylines to present the material available in ways that would be appealing to television audiences. If he identified an arresting sequence upon which a story could be constructed, he would ask for more material. For example, coming across a sequence depicting a species of wasp, Parsons found it 'quite obvious that Ammophila will turn into a major sequence that can be run at some length', and therefore requested 'a few build-up shots'.

> I suggest that you show one wide angle establishing scene of an observer, which could be either Eric or yourself, walking up and looking at the nest site with an additional mid-shot cut away and an additional scene to show the relation between the nest site and the area from which the mud was gathered. The last shot might possibly be done with a pan or walking the observer from one side to the other.[19]

Through these instructions, Parsons reminded the Oxford film-makers of the grammar of visual storytelling in television programmes. But as this material was also to serve for the production of visual teaching aids of various kinds, Parsons's guidance turned him into a participant in the production of Thompson's educational films.

On the whole, Parsons did not get from the expedition as much material as he had hoped, especially when it came to Thompson's and Skinner's trademark macro-cinematography of insects. However, Parsons judged Peter Parks's material on plankton to be 'first class'. This is perhaps, he wrote to Crocker, 'the biggest breakthrough in nature cine photography we have had for some time'.[20] Parks's film material was in fact 'more or less based on Sir Alister Hardy's book "The Open Sea: The World of

---

[18]Christopher Parsons, personal letter to Gerald Thompson, 15 August 1967, BBCWAC WE13/1,071/1.

[19]Christopher Parsons, personal letter to Gerald Thompson, 6 September 1967, BBCWAC WE13/1,071/1.

[20]Christopher Parsons, 'Thompson Jamaica Expedition', memo to Editor, Natural History Unit, 25 October 1967, BBCWAC WE13/1,071/1.

Plankton"', and so Parsons suggested that one entire programme should be devoted to presenting it. This programme idea became *The Living Sea* (1968), broadcast as an episode of *The World About Us*. The rest of the best footage Thompson and his colleagues had brought back from Jamaica was presented in *Wild Jamaica*, as two successive episodes of *The World About Us* on the marine and terrestrial life of the island. Although the expedition and its output did not quite meet Parsons's expectations, the Jamaica project established the idea that the NHU could not avoid working with this group of film-makers who would soon be trading as OSF.

The Making of a Natural History Film: Wildlife Film-Making as Augmented Reality

Writing in March 1969 to the head of the NHU, Nicholas Crocker, Parsons intimated, 'I think the time has come for us to seriously consider some sort of contract with them [OSF] which will enable us to dupe off original material as it is shot and before it is delivered to companies like the Ealing Corporation.'[21] Upon returning from Jamaica, Gerald Thompson had approached this American company specialising in the production and distribution of educational films, and secured £120,000, which in effect guaranteed the existence of OSF for the following five years and its independence from Oxford University. Parsons was convinced that OSF's 'highly specialised and, in some cases, unique material' could be of interest across BBC Television. In his view, not only natural history programmes from Bristol but also the science programme *Horizon*, produced in London, or the various programmes of Schools Television, could have an interest in securing access to content from OSF. He thus suggested that the BBC guarantee an annual payment to the Oxford unit to secure priority access to all their material. However, no bureaucratic mechanism existed at the BBC to allow this to happen, and so Crocker recommended to Parsons that he should enlist David Attenborough's support:

---

[21]Christopher Parsons, 'Oxford Scientific Films Ltd.', memo to Editor, Natural History Unit, 12 March 1969, BBCWAC WE21/50/1.

I think the only hope is for you to have a word with David if you can. Otherwise I think the whole thing might result in trying to get money from about twenty different programme sources, in which case we might end up with almost nothing at all.[22]

The paper trail in the BBC archives ends there, not revealing whether Parsons's idea materialised. But soon after Parsons had suggested that different departments at the BBC should work with OSF, Mick Rhodes (1934–2018), a producer working for the BBC2 science series *Horizon*, began taking an interest in the Oxford film unit. This collaboration would eventually lead him from London to Bristol, where he became head of the NHU, profoundly reshaping the unit and inflecting its coverage of natural history topics toward a more scientific approach.

With Mick Rhodes, the head of the Bristol unit at last combined training as a scientist with the skills of a seasoned television producer. Before working for television, he had read Zoology at Bangor University and had been a practicing ecologist for a couple of years, specialising in the study of populations of bluebottle flies. He then began, in 1965, a career at the BBC, first producing science radio programmes and then moving on to become a television assistant producer in the Science and Features Department, starting work on the *Horizon* series in 1968.[23] In 1969, Rhodes invited OSF to participate in his project to film, over the course of a year, the work of zoologists investigating the ecology of Wytham Wood. This venture would eventually become the *Horizon* programme *The Wood*, broadcast in February 1971. Located on a hill above the Thames near Oxford, the thousand acres of Wytham Wood had been bequeathed to Oxford University in 1943 under the condition that they would be preserved and used as teaching and research facilities. Since then, at any given time, more than twenty scientists had been constantly observing, counting, measuring, or marking representatives of the 4000 species living there, trying to unravel the webs of ecological relationships between

---

[22] Editor, NHU, 'Oxford Scientific Films Ltd.', memo to Chris Parsons, 13 March 1969, BBCWAC WE21/50/1.

[23] Peter Goodchild, 2018, 'Mick Rhodes obituary', *The Guardian*, 5 November 2018. Available online at https://www.theguardian.com/tv-and-radio/2018/nov/05/mick-rhodes-obituary. Last accessed 15 January 2019.

them. Wytham Wood, as Mick Rhodes remarked, was 'the most studied and most understood wood in the world'.[24] Starting in February 1970, Rhodes began filming the researchers at work, observing, capturing and analysing specimens. He stopped in February 1971. In the meantime, he contracted OSF to film up-close some of the life forms scientists were studying, such as oak-eating caterpillars and their predators: shrews, beetles, owls, and great tits. The camera peered into nest boxes and filmed insects and their larvae up close. The footage offered views of these life forms and their life cycle from a perspective that could pass for that of the scientists studying them. OSF's film-makers thus offered television viewers the impression that they shared in the scientists' expert gaze.

Rhodes's collaboration with OSF for *The Wood* was the beginning of an enduring relationship whose most visible and tangible outcome turned out to be *The Making of a Natural History Film*, another *Horizon* episode broadcast in November 1972 and this time depicting the filming methods deployed in the Oxford film unit.[25] *The Making of a Natural History Film* is the second wildlife making-of documentary produced in Britain for television. Just like the 1963 *Unarmed Hunters* (Chapter 5), it portrays a wildlife film unit at work and, in doing so, lays claims about what the standard of wildlife film-making should be. It affirms the shift toward the precedence of the mastery of the film-making apparatus over the skills of the field naturalist which underpinned the professionalisation of wildlife television production started in the early 1960s. It also signals that, by the early 1970s, the displacement of natural history by zoology as the intellectual foundation of wildlife television production in Bristol, engaged in 1965, was complete.

The idea of *The Making of a Natural History Film* had come to Mick Rhodes when watching the Oxford film-makers at work for *The Wood*. The two features are similar in that they both show a group of people at work to produce knowledge of nature, in the latter case, wildlife film-makers. For the purpose of the documentary, Parks and his colleagues agreed to produce a film on one of the main animal models in science,

---

[24] Mick Rhodes, 1971, 'The Wood', *The Listener*, 85(2200), 665–668, 668.

[25] *The Making of a Natural History Film* went on to win the Italia Prize in 1973, the same award Tinbergen's *Signals for Survival* had won in 1969, and got repeated many times on the BBC.

the stickleback fish. Rhodes and his team would in turn film them as they were tackling the various practical problems involved with a camera in exploring the fish's mating behaviour. Rhodes, who himself had worked in the field, had intended with *The Wood* to dispel the notion that ecologists were an 'inferior' sort of scientist because they worked in the field and used what looked like low-tech methods. What mattered ultimately were their motivations and the impact their work had on everyone's daily life. As Rhodes put it, although ecologists' 'equipment is often a Heath Robinson affair of string', their 'values may do more to allow us to keep our world bearable than all the particle-accelerators and physicists in their expensive Dr Who film-sets'.[26] Likewise, Rhodes's portrayal of the Oxford wildlife film-makers cast film-making as a legitimate method of investigating nature, the film depicting the convoluted equipment Parks and his colleagues used to obtain such deceptively simple shots as that of a fish spawning on a river bed.

In its first part, the film presents the different techniques and equipment devised and used at OSF. The second part is a making-of documentary for the stickleback feature. Film-making is presented here as a succession of problems of representation to be solved, using zoological knowledge and technical ingenuity so that living beings behave in front of the camera according to expectations. The filming space, where many creatures (bats, owls, mice, and various plants) live freely, appears cluttered with water tanks, sets reproducing a section of a meadow with living plants, and a complex looking optical bench. The film-makers shown at work there emerge as inventors, deploying their film craft to devise ingenious devices that help them create representations of the natural world from surprising angles and thus enrich viewers' perception, enabling them to witness the functioning of nature from perspectives impossible to get from a simple stroll in the wild. In one visually remarkable sequence, Peter Parks is shown preparing a pitcher plant to film the capture of a fly. The originality of the shot is to appear as if taken from within the pitcher plant itself. To obtain this unusual point of view, Parks cut in half the cavity formed by a cupped leaf of a pitcher plant. He then glued the upper half onto a glass plate, filling it with the liquid that attracts insects in the plant.

---

[26]Mick Rhodes, 1971, 'The Wood', *The Listener*, 85(2200), 665–668, 668.

Positioning the whole above a mirror inclined at 45 degrees and pointing a camera at it captured the view one would have, had the camera been placed at the bottom of the pitcher. The original device depicted on-screen demonstrates the value of intervention to obtain truthful representations revealing facets of nature which would otherwise be impossible to see. The sequence shows the film-maker taking apart nature only to re-assemble it on-screen. It delegates the act of observation to the film-making apparatus and claims a place for this interventionist approach to wildlife film-making within the scientific sphere, demonstrating OSF film-makers' freedom to distance themselves from nature in order to reach at what is presented as an objective truth about it (Daston and Galison 2007). In *The Making of a Natural History Film*, naturalists' belief in non-intervention and self-effacement is repudiated. The documentary celebrates film-maker's active, hands-on, physical engagement with the natural world as a legitimate way of getting to know it.

The film-makers' initial scientific training is their key feature, high-lighted throughout the documentary. From the outset, the members of OSF are defined in the voice-over commentary as 'a group of ex-university zoologists who have given up their dreaming spires to become full-time natural history film-makers'.[27] Although this opening commentary presents as natural the distinction Robert Reid was asserting in 1969 between scientists and film-makers, claiming that one could not be both at the same time, Parks's and his colleagues' initial scientific training is at the same time presented as what frees them from the natural history constraint of non-intrusion, liberating their creativity and their ingenuity when it comes to deploying innovative filming techniques. The camera no longer simply mediates between the naturalist cameraman and the audience so that the latter can partake in the former's expert encounter with the wild, as in a *Look* episode for instance. Instead the camera reveals the natural world as neither the film-maker nor the audience could ever see it, save for the interposition of the camera between nature and the observer. Likewise, OSF film-makers' field expeditions are shown as the antithesis of an amateur naturalist cameraman's venture in the field. In the 1963 *Unarmed*

---

[27] *The Making of Natural History Film*, BBC, Transmission date: 23 November 1972.

Hunters, Eric Ashby appeared progressing with precaution, keeping himself silent and concealed to avoid alerting the animals he was trying to film. By contrast, an OSF party's excursion in the field, mostly intended to capture specimens to bring back to the studio, is a boisterous occasion, far from the field naturalists' ethos of self-effacement. Highlighting the importance of OSF's filming equipment and techniques, *The Making of a Natural History Film* turned wildlife film-making into a performance of knowledge production, an equal to field scientists' work when it comes to producing knowledge of the natural world.

The Making of a Natural History Film also had profound consequences for the production of wildlife television in Bristol. The film went on to win a BAFTA[28] for best factual programme, and a little later, the Italia Prize for best documentary. Even before this episode of *Horizon* had been broadcast, David Attenborough, then the director of programmes at the BBC, had appointed Mick Rhodes to the position of head of the NHU in Bristol. The decision came as a shock to the producers in Bristol, where it was widely believed that Christopher Parsons would succeed Nicholas Crocker (Parsons 1982). During his last appraisal interview in March 1972, when asked for his views on who should replace him, Crocker had shared his opinion that 'Chris Parsons was now mature and experienced enough to make a good job of it.' Parsons had good standing with the NHU's outside contributors and contacts, as well as excellent relations with the staff. Crocker went on to praise Parsons's organisational skills.[29] In a later interview, Parsons recalled having been disappointed with not getting the position. At the same time, 'There was a lot of dead wood in the Unit, and it needed a shake-up, and it would have been quite difficult for me to do, within the Unit.'[30] Attenborough chaired the board which appointed Rhodes. His intention in installing at Bristol a science television producer trained on *Horizon* was to import to the south-west region the methods and the approach to science that prevailed in the London-based

---

[28] British Academy of Film and Television Arts.

[29] Record of annual interview, 1 March 1972, BBCWAC L1/2,264/1.

[30] Christopher Parsons, 2001, Oral history interview, Wildscreen.

Science Unit. The aim was to strengthen the links between the NHU and the sciences and change the way producers in the NHU operated.[31]

Rhodes ruffled some feathers but reformed practices at the NHU, enticing the people working there to think anew of old subjects. Exemplary of his influence are a couple of programmes both intended to renew the presentation of badgers, a staple of wildlife television from the early *Look* programmes onward. In 1975, Rhodes convinced Ashby, who considered filming habituated animals as a deception of viewers, to construct an artificial badger sett to obtain intimate footage of badgers' family life. Connecting drain pipes with concrete, Ashby created an artificial tunnel leading to two nest chambers installed in one half of a woodland shed. A sheet of glass concealed Ashby's presence in the other half of the shed. After several months spent first waiting for badgers to discover the place and then habituating them to the lighting in the shed, Ashby obtained the first-ever footage of badger behaviour taken from inside a sett (Bale 1982: 18). With this film, which took him more than three years to make, Ashby had adopted the interventionist approach favoured in Oxford, based on filming under controlled conditions. *At Home with Badgers*, broadcast in the *Wildlife on One* series in 1978, narrated by David Attenborough, belongs in the same category of wildlife documentaries as Sielmann's *Woodpeckers* (1955) and Eastman's *Private Life of the Kingfisher* (1966). All three films rest on the construction of artificial filming environments to offer viewers the spectacle of wild animals' domestic life.

As Ashby was busy figuring out how to best light the inside of his woodland shed, and anxiously waiting for badgers to visit his man-made sett, Rhodes came up with another programme idea. This was to be a live transmission from outside a badger sett, each evening for a week, for ten minutes. Key to the project were remotely controlled infrared cameras, a technology developed for the military. As Rhodes explained to Ashby, these cameras that could 'shoot in total darkness', were also 'completely remote. They can pan, tilt and zoom at the press of a distant button and make very little noise as well'.[32] Rhodes was querying Ashby, as an expert on badger matters, about what would be the best time of year

---

[31] Interview with author, 25 January 2016.
[32] Mick Rhodes, letter to Eric Ashby, 11 August 1976, p. 1. BBCWAC WE13/541/1.

for the broadcast, what sort of typical activity the camera was likely to capture, or if badgers were 'reliable' and could be expected to perform typical behaviour at the same time in the evening every day of the week.[33] Ashby resisted Rhodes's efforts at formalising badger behaviour, arguing that there was much that was unpredictable there, noting: 'Perhaps people badger-watch because there is no such thing as a typical evening—it's different every time!'[34] Eventually, Peter Bale, who was to produce *Badgerwatch* (1977), approached another badger expert, Ernest Neal, who put him in touch with Dr Chris Cheeseman from the Ministry of Agriculture, 'who was doing fieldwork in Gloucestershire with the controversial "badger-gassing" programme' (Bale 1982: 11). On 9 May 1977, the 'first badger ever seen live on air' (Bale 1982: 15) appeared on television. With *Badgerwatch*, the NHU renewed the use of outside broadcasting, which had encapsulated the essence of television before the studio and later pre-recorded films took over. By going live, despite the 'vast complication of the equipment and people involved'[35] the Bristol broadcasters invited audiences to practice observational natural history in their living room through watching television, reviving the impulse at the root of post-war wildlife television as exemplified in the *Look* series (Chapter 2).

But at the same time, *Badgerwatch* presented itself as a technologically enhanced version of the first *Look* series, thus appearing to perfect it, in accordance with the deterministic narrative of technical progress which pervades the public presentation of wildlife film-making. This programme also renewed the meaning attached to the technique of live outside broadcasting, which, in the 1930s or 1950s, epitomised the 'unmediated mediation' of television. It enabled viewers to remotely participate in distant events taking place independently from the presence of cameras (e.g. the coronation, or a sporting event). In *Badgerwatch*, filming creates the event. Here, outside broadcasting is enrolled into an endeavour whereby the event broadcast live only becomes part of audiences' lived experience because it is filmed. If it were not filmed with these automated, remote-controlled cameras, nobody would be there to observe the badgers. In the

---

[33] Mick Rhodes, personal letter to Eric Ashby, 15 July 1976, BBCWAC WE13/541/1.
[34] Eric Ashby, personal letter to Mick Rhodes, 25 July 1976, BBCWAC WE13/541/1.
[35] Mick Rhodes, letter to Eric Ashby, 11 August 1976, p. 2. BBC WAC WE13/541/1.

spirit of OSF's superseding field craft with film craft, the skills of the amateur naturalist cameraman, like Eric Ashby, who had made a specialism of patiently waiting to film badgers, were delegated to a technological device. Wildlife seen through the camera supersedes any unmediated encounter with nature.

Mick Rhodes's arrival as editor of the NHU started a transfer of competences from the BBC science department, in London, to the Bristol unit, several *Horizon* producers moving to the south-west region. Michael Andrews moved to the NHU in 1976 to become the Bristol-based editor of *The World About Us*. His London counterpart was another former *Horizon* producer, Anthony Isaacs. Peter Jones, who likewise had started his BBC career on *Horizon*, made the move to Bristol in 1979 to replace Andrews, later becoming editor of the series *The Natural World* (1983–). This renewal of personnel in Bristol also ascertained, as much as it sustained, the move away from amateur natural history toward the sciences as a new framework of reference for the NHU's output. Rhodes's appointment became effective on 23 October 1972. One month later, David Attenborough announced his resignation from his position as BBC Television's director of programmes to become a freelance programme maker. The news came as a surprise to most at the BBC, as 'no hint of his departure had leaked out'.[36] *The Making of a Natural History Film* was broadcast two days later, on 26 November, signalling to audiences that the new direction of travel was the production of programmes whose intellectual foundations would be found in zoology much more than in traditional amateur natural history. OSF was the standard bearer of this renewed approach, revolving around a conception of film-making as a means of augmenting audiences' perception and understanding of the natural world.

Yet, in the meantime, the relationship between Oxford and Bristol had severely deteriorated to the extent that most of the producers in Bristol were refusing to work with the Oxford film-makers, 'throw[ing] up [their] hands in despair at the prospect of dealing with them'.[37] This situation had arisen from OSF film-makers punctiliously disputing the conditions set in each of their contracts with the BBC so as to always extract a supplementary

---

[36]Chris Dunkley, 'BBC TV programmes head quits', *The Times*, 24 November 1972, p. 3.

[37]Mick Rhodes, 'Payment to OSF', memo, 27 November 1972, BBCWAC WE13/268/1.

payment from the Corporation. One of Rhodes's first actions as the new head of the NHU was to step into defuse a situation threatening to result in the break-up of the relationship between OSF and the NHU: 'In the future I think we shall need OSF's skills and therefore I'm at pains to bury all hatchets and start again, but with care and no softness: We do not owe them a living.' [38] OSF, for their part, had expressed to Rhodes their wish for a 'fresh start', and 'Gerald Thompson has invited Jeffrey [sic.] to visit them, so it all looks good'. To ensure that the relationship would work, Rhodes established himself as a go-between:

> Because our relationships with OSF have been confuse in the past, and because they and I am keen to keep them unconfused in the future, will anybody contacting them in a way that could either lead to a contract or to them doing work on our behalf, for which they might expect to be paid, please talk to me first.[39]

From this point onward, the NHU in Bristol took a scientific turn under the editorship of Mick Rhodes, whose ambition was to produce programmes combining 'entertainment-with-information with education'.[40] This approach would materialise most substantially by the end of the decade in the first major natural history series undertaken in Bristol, featuring David Attenborough and *Life on Earth*, and in which OSF contributed key footage.

Conclusion

The relationship between OSF and the NHU was a dialectical one. As much as the NHU pro-actively encouraged OSF to grow and develop, contracting them as often as they could, OSF exerted a transformative influence on the Bristol unit. In the process, wildlife television also changed. Taking their cues from lab work rather than field natural history, the

---

[38] Mick Rhodes, 'Payment to OSF', memo, 27 November 1972, BBCWAC WE13/268/1.
[39] Mick Rhodes, 'Payment to OSF', memo to All NHU production staff, 28 November 1972, BBCWAC WE21/50/1.
[40] Mick Rhodes, personal letter to David Attenborough, 27 October 1972, BBCWAC WE8/83/1.

Oxford film-makers devised new methods for, and a new understanding of, filming wildlife. They brought nature into the film studio instead of taking their cameras into the field, making intervention and artifice a prerequisite of the production of visual representations of nature. In other words, they turned on its head the key tenet of the culture of wildlife film-making as it had developed from the culture of amateur natural history—'Thou shall not intervene'—which shunned, as falsification and deception, physical interaction with nature when filming. With OSF, film-making became an active means of investigating the natural world rather than a way of passively recording observations. Instead of making the knowledge value of film footage conditional on a negation of the encounter between the camera and its subject, the Oxford cameramen made this encounter the key moment of knowledge production, which audiences were invited to witness and celebrate. The making of the film became a testimony to the film-makers' skills at manipulating both nature and the camera.

The key idea governing OSF's modus operandi is that the camera, rather than being a passive tool, is an active instrument. The purpose of filming is not to take viewers to the field with the cameraman but to invite them to share in the cameraman's quest to see nature through the camera. Camerawork becomes a key aspect of the narrative in wildlife films, prompting audiences to wonder how the footage could be obtained. This approach was in harmony with the ethos informing the actions and beliefs of broadcasters at the BBC in the late 1960s–early 1970s, centred on the notion of professionalism, where the main concern is with the technical quality of the performance, against set standards of appraisal. Here, broadcasting, the production of television programme of high technical quality, becomes an end in itself (Burns 1977: 128). OSF cameramen's main preoccupation with producing filmic tours de force not only fitted in this ethos but helped reinforce it. Introducing the notion of technical progress in wildlife film-making for all to see, the films produced in Oxford helped make earlier wildlife film-making obsolete. As the final chapter considers, this evolution contributed to making *Life on Earth* and the following 'Attenborough series' possible and made the adjunction of making-of documentaries to these blockbusters a necessity.

References

Bale, P. (1982). *Wildlife through the camera*. London: British Broadcasting Corporation.

Burkhardt, R. W. (2005). *Patterns of behavior: Konrad Lorenz, Niko Tinbergen, and the founding of ethology*. Chicago and London: The University of Chicago Press.

Burns, T. (1977). *The BBC: Public institution and private world*. London: The Macmillan Press.

Canales, J. (2011). Desired machines: Cinema and the world in its own image. *Science in Context, 24*(3), 329–359.

Crowson, P. S. (1981). *Animals in focus: The business life of a natural history film unit*. Horsham: Caliban Books.

Curtis, S. (2015). *The shape of spectatorship: Art, science, and early cinema in Germany*. New York: Columbia University Press.

Dagognet, F. (1992). *Etienne-Jules Marey. A passion for the trace*. New York: Zone Books.

Daston, L., & Galison, P. (2007). *Objectivity*. New York: Zone Books.

Parsons, C. (1982). *True to nature*. Cambridge: Patrick Stephens Ltd.

Shaffer, L. (1991). The Tinbergen legacy in photography and film. In M. S. Dawkins, T. Halliday, & R. Dawkins (Eds.), *The Tinbergen legacy* (pp. 129–138). London: Chapman & Hall.

Thompson, G., et al. (1981). *Focus on nature*. London: Faber & aber.

9

Life on Earth and Beyond: Producing the Wildlife Blockbuster

In autumn 1972 a new epoch began for the BBC NHU and British wildlife television, one in which we still find ourselves today. The Unit had a new head, Mick Rhodes, convinced of the necessity to integrate wildlife television more closely with the scientific exploration of the natural world. Producers at Bristol had got past trying to figure out how to produce wildlife television programmes and were now operating within a well-defined professional framework. A web of professional wildlife cameramen, nurtured by the NHU throughout the latter part of the 1960s, was active around the world, providing the Unit with the footage needed to keep renewing its visual offer to British television audiences. The notion that the broadcasting of wildlife programmes is a process of television, under the professional direction of television producers, not naturalists or field scientists, had become evident. Finally, as an internationally recognised centre of expertise in the production of wildlife films, the NHU was well established in the scientific world as a trusted interlocutor. The Unit had access to an international network of scientists in research institutions, ready to advise producers preparing programmes on different aspects of life on Earth. It is in this promising context that David Attenborough

© The Author(s) 2019
J.-B. Gouyon, *BBC Wildlife Documentaries in the Age of Attenborough*,
Palgrave Studies in Science and Popular Culture,
https://doi.org/10.1007/978-3-030-19982-1_9

announced to a stunned BBC his resignation from a position of great power within the BBC, that of director of television programming. The time was ripe for him to become a freelance programme maker.

'The Time Has Come for a Large-Scale Natural History Series'

Just a few months before Attenborough's resignation, in August 1972, two producers at Bristol, Christopher Parsons and John Sparks, independently expressed their conviction that the time had come for the NHU to commit resources and personnel to the production of a large-scale natural history series of the same kind and proportion as the 1969 *Civilization*. The two proposals shared their analysis of where wildlife television in Britain stood in 1972 and how it should evolve, but they differed in their appreciation of the cultural status of wildlife and natural history. John Sparks's proposal was for 'a wildlife version of Kenneth Clark's "Civilization"'. In seven to thirteen episodes, each being thirty or fifty minutes long, the aim of the series would be to portray 'European man's discovery of the world's animal and plant communities', and to examine 'the effect that new discoveries had upon people'.[1] To Sparks, this series 'should be offered to BBC-1 because, unlike "Civilization" or "The Ascent of Man", animals (wildlife—call them what you like) are intrinsically more popular with the broader range of people than culture'.[2] Sparks's assessment, however, was going against the will to make wildlife a part of culture, which had been the guiding principle of all new developments in wildlife television since Attenborough had arrived at BBC2 in 1965. In the early 1970s, natural history had all but disappeared from BBC1's schedule. Bristol's flagship series *Look* had ended in 1968. The new standard bearer of wildlife television, *The World About Us*, started in December 1967, was broadcast on

---

[1] John Sparks, 'Major natural history series', memo to Editor, NHU [N. Crocker], 8 August 1972, BBCWAC WE17/53/1.

[2] John Sparks to Editor NHU [N. Crocker], 'Major natural history series', 8 August 1972, BBCWAC WE17/53/1.

BBC2, a channel dedicated to offering 'programmes of real cultural significance'[3]. As soon as he had been appointed controller of BBC2, in 1965, Attenborough endeavoured to renew wildlife television and raise its cultural profile, moving it closer to professional science, away from amateur natural history. The series *Life*, which turned into a mainstream one the relatively heterodox idea that insights about human nature could originate in the scientific study of animal behaviour, played a key role in this project (Chapter 6). BBC2 became the source of new wildlife programme ideas, which were then imported to BBC1.

Christopher Parsons, by then a senior wildlife broadcaster and more attuned than his younger colleague to the dynamics prevailing in his field, addressed his memo directly to Robin Scott, Attenborough's successor as controller of BBC2, with a copy to Nicholas Crocker, the outgoing editor of the NHU. In the explanatory note accompanying the proposal, Parsons indicated his intention to talk to Attenborough to gauge his reaction to the idea for the series, with the notion that he could be involved as a presenter for what was already titled *Life on Earth*.[4] More elaborate than Sparks's proposal, Parsons's twenty-part proposal also took an approach much closer to the usual themes of the programmes coming out of Bristol. Taking the perspective of scientists' work on the topic, principally 'ethologists, palaeontologists, biochemists or whatever', the series was to 'tell—as far as present-day knowledge goes—the story of life on this planet from its appearance about two and a half thousand million years ago, through the continuing process of evolution, until now'.[5] Each fifty-minute episode was to concentrate on one topic and be self-contained, 'capable of being intelligently and entertainingly viewed in itself'. Parsons's proposal conceived of *Life on Earth* from the outset as an international co-production project. The increased budget this approach permitted made ambitious filming possible and guaranteed an overall high production value for the series. The scripts were to be 'backed by a team of experts', and the series

3 'Programme content of second channel: Summary of statements made up to now', 19 December 1960, BBCWAC T16/538. Quoted in Boon (2017).
4 Christopher Parsons, '"Life on Earth": Outline proposal for major natural history series', memo to Controller, BBC-2 [Robin Scott], 24 August 1972, BBCWAC T41/520/1; Anonymous, n.d., 'Life on Earth. Notes from file on what happened …', p. 1. BBCWAC WE17/53/1.
5 Christopher Parsons, 'Life on Earth. Outline proposal…', p. 2. BBCWAC T41/520/1.

was to be presented 'by two personalities, both experienced television performers as well as zoologists and anthropologists in their own right'. Here, Parsons had a very precise idea of who these two presenters should be: Desmond Morris and David Attenborough.[6] They would 'leapfrog' from one key location to the other as the story of each episode unfolded. Mindful of the constraints attached to international co-production projects as well as to the value of being able to sell the series abroad, Parsons envisaged that presenters' on-screen appearances would be 'filmed in such a way that they can be replaced by filmed presenters in other languages'. Presenters were not to be the focal point of interest: 'Each film will be so pictorially riveting that it provides entertainment at a visual level alone.'[7]

'Chris Parsons's project',[8] as it was referred to in internal correspondence, received the support of Stuart Wyton, regional senior manager in Bristol, who weighed in with the channel controller. However, Robin Scott refused to table the proposal at the March 1973 commissioning meeting, querying the notion that the series should use two presenters and be in twenty parts. To Scott, the model to follow was Jacob Bronowski's *The Ascent of Man*, about to be broadcast, a thirteen-part programme, with one presenter. Parsons nonetheless went to find out whether Attenborough would be interested in fronting the series. The conversation between the two men, who'd known each other for more than a decade, was positive. It convinced Attenborough that the time had come to resign: 'When I resigned, what I wanted to do was *Life on Earth*.'[9] Before leaving, Attenborough appointed Mick Rhodes as the new head of the NHU and met with Michael Peacock, his former colleague at the BBC, who had gone to America and was now the executive vice-president in charge of programmes of Warner Brothers, the American television network. Peacock showed immediate interest in the *Life on Earth* idea.[10] The series was not yet commissioned, but the gears had been set in motion.

[6]Stuart Wyton, 'Life on Earth', letter to Robin Scott, 24 August 1972, BBCWAC T41/520/1.

[7]Christopher Parsons, 'Life on Earth. Outline proposal…', p. 2. BBCWAC T41/520/1.

[8]Stuart Wyton, personal letter to Robin Scott, 24 August 1972, BBCWAC T41/520/1.

[9]Interview with author, 25 November 2015.

[10]Anonymous, n.d., 'Life on Earth. Notes from file on what happened …', p. 1. BBCWAC WE17/53/1.

Planning for the new series began immediately after Attenborough's resignation had become effective. In January 1973, he travelled to Bristol for two days to record the commentaries for two episodes of the series *The World About Us*. On his second day there, a meeting took place with Mick Rhodes, Christopher Parsons, and John Sparks, lasting a whole afternoon, to get the 'Epic series' off the ground.[11] According to Rhodes, they 'reached a surprising level of agreement', hatching 'an outline division of the subject into programmes' and coming up with some form of staffing arrangement for what was to be 'the definative [sic.] series on how man got to where he is in the biological sense; a history not of his ideas but of his body and behaviour—as somebody put it, "the Decent [sic.] to man"'. Parsons, Attenborough and Sparks, who 'seem to nicely counter point each other' would form a 'triumvirate with David holding the casting vote as executive editor'. In addition, Attenborough was to be 'the storyteller' throughout, so as to ensure a 'constancy of style', and because 'judging by the material back from Indonesia', Rhodes found him 'most attractive in the performance sense'.[12] In the course of the conversation Attenborough mentioned his meeting with Michael Peacock, and Mick Rhodes met with him in London a week later. Peacock wrote quickly afterward to confirm his interest in co-producing *Life on Earth*, hoping for a transmission in the autumn 1975. At this early stage, all those involved were 'madly enthusiastic about this project'.[13] Yet, after agreeing on the basic principles for the series, almost a year would go by before anymore planning work took place. Before *Life on Earth*, Attenborough—who had never really left the limelight—needed nonetheless to reconnect with his public. When still the director of television programming at the BBC, Attenborough had been plotting, in relative secrecy, the wildlife television programme which would revive his relationship with television viewers and mark his comeback as a wildlife presenter. *Eastward with Attenborough* was a spin-off of the *Zoo Quest* concept, set in the land of his past successes, Indonesia. Over three months,

---

[11] Anonymous [probably M. Rhodes], 16 January 1973, Epic Series—Discussion with David Attenborough, John Sparks, Chris Parsons and Mick Rhodes. BBCWAC WE17/53/1.

[12] Anonymous [Mick Rhodes], n.d., 'Life on Earth, a proposal for a major series from the NHU', p. 1. BBCWAC T41/520/1.

[13] Christopher Parsons, '"Life on Earth"—A major natural history series', memo to Controller BBC-2 [Robin Scott], 4 October 1973, BBCWAC T41/520/1.

Attenborough, his producer and a film crew travelled from Borneo to Sumatra through to the Celebes and Java, filming flying snakes, locals preparing and eating bird's nest soup, the Krakatoa, and a host of animals along the way. To justify going to Indonesia, Attenborough used a similar argument to the one he used for his first *Zoo Quest* series. Speaking to *Radio Times* reporter Liz Cowley, he explained that although viewers were getting 'a glut of wild-life film', it was all from Africa. By contrast, South-East Asia was a 'relatively untapped' area.[14] To assist him, Attenborough had chosen one of the assistant producers trained on Desmond Morris's *Life* series, Richard Brock: 'And so I was appointed for some reason, because I was just an assistant producer, to be a producer with the great David Attenborough who was number one in the 24,000 staffed BBC.'[15] As Brock remembered, this involved regular visits to Television Centre in London, 'this big circular building and you kind of walk across the middle'.

> He was, of course, keeping all this secret. I'd go in and we'd get out the map of Borneo …, and start planning this. … We went and we were there for three months with an excellent crew, three people, and we made a series called *Eastwards with Attenborough*. So that was his comeback really after *Zoo Quest*. He'd gone into the BBC from *Zoo Quest* and then reappeared on *Eastwards with Attenborough*.[16]

With this series, the wildlife presenter wanted to claim his crown back after years of self-imposed exile in the corridors of power. The series was well received, as much within as outside the NHU. Each episode was broadcast twice, on a Thursday, with a repeat on the following Monday, in October and November 1973, therefore maximising viewers' exposure. Writing as the transmission of the series was nearly complete, Richard Brock congratulated Attenborough: 'I've just come from a Natural History Unit meeting at which the series was warmly acclaimed. Your performance and the high technical quality were very much appreciated … I am sure you are totally immersed in a golden haze.'[17] Television critics were likewise

[14]Liz Cowley, 'Attenborough opens his Jungle Book', *Radio Times*, 4 October 1973, pp. 6–7.

[15]Richard Brock, 2007, Oral history interview. Wildscreen.

[16]Richard Brock, 2007, Oral history interview. Wildscreen.

[17]Richard Brock, personal letter to David Attenborough, 6 November 1973, BBCWAC WE8/83/1.

full of praise and willingly took part in the chorus celebrating the return of the 'wandering naturalist'.[18] The series, 'done Bristol fashion with an admirable crew and with Richard Brock as producer', Leonard Buckley noted in *The Times* on the morning following the transmission of the first episode, 'firmly and finely rekindled' viewers' taste 'for Mr Attenborough's special sort of travelogue'. *Eastward with Attenborough* brought the prodigal son of British wildlife television home: '"Isn't it nice," whispers somebody outside the cutting room, "to have young David back where he belongs—wading through all those ropes of film about flying squirrels and mud-skippers and things?"'.[19] Pointed references to *Zoo Quest* throughout the series, both in terms of setting (Indonesia) and style (the travelogue), suggested that Attenborough was picking up wildlife television where he had left it a decade earlier. As the double spread richly adorned with colour photographs of wildlife in the *Radio Times* suggested, little had changed in the interval. The main difference, which to an extent justified the series, was the introduction—in the meantime—of colour television. With this new series, Attenborough asserted his primacy over British wildlife television. Meanwhile, Christopher Parsons had kept thinking about *Life on Earth*. Whilst Attenborough was busy filming around Indonesia and then editing the footage, Parsons was reappraising some early assumptions, both conceptual and organisational. Notably, he had become worried that the programme devoted too much time 'to Man' and insisted 'It is important that we get man in the right perspective and this will obviously need further discussion'.[20] Besides, the scale of the project was starting to dawn on the NHU. Confronted with the hiatus *Eastward* had introduced in the planning, Parsons suggested that between Attenborough's other commitments, and the time needed to recruit the project's team, Mick Rhodes might have been a little too optimistic about the proposed working arrangement for the series. Attenborough's personal projects—not only appearing in other programmes but also writing books and delivering public lectures—made it unrealistic for him to be based at Bristol as a member of staff of the NHU for two to three years. Parsons suggested limiting his involvement

[18] Leonard Buckley, 'Eastward with Attenborough, BBC1', *The Times*, 12 October 1973, p. 15.

[19] Liz Cowley, 'Attenborough opens his Jungle Book', *Radio Times*, 4 October 1973, pp. 6–7, 6.

[20] Christopher Parsons, '"Life on Earth"—A major natural history series', memo to Controller BBC2, 4 October 1973, BBCWAC T41/520/1.

to the planning stage, 'when we decide how many programmes there will be, the subject material to be included in each and the styling of the series as a whole', followed with an intense period of round-the-world location filming, to conclude with 'commentary writing and recording' at the end of the production period. Although *Life on Earth* was turning into the NHU's main priority, it was not yet tangible enough for Attenborough to commit to it. For, having just gone freelance, he was eager to re-establish a strong presence as a wildlife television presenter, and so concentrated on other projects. For a short while, the planning for *Life on Earth* entered a period of uncertainty.

'A Programme Undertaking of Unusual Dimensions'

But the arrival of Aubrey Singer as the new controller of BBC2 in January 1974 imposed a new dynamic on the project. Singer notably insisted early on that *Life on Earth* should become Attenborough's priority. The creator of *Horizon* took from the outset a hands-on approach and was decisive in shaping what was to be the most important television programme produced under his oversight as the head of BBC2. Singer and Rhodes knew each other very well from their common past involvement with *Horizon*. This precedent made it easier for Singer to get closely involved with the production of *Life on Earth*. And eventually, it reached the top of Attenborough's agenda. His treatment for the programme, the indispensable first step in providing a vision and a sense of the story for the series to come, began circulating in the BBC in September 1975. The producers—Parsons, Parks, and Brock—were to use it as the basis from which they could start planning sequences to film and begin researching them. It was also sent to potential co-producers, such as Michael Peacock at Warner Bros, to initiate negotiations. A potential financial backer—he offered to contribute £25,000 per episode for a thirteen-part series—and a former BBC executive who'd worked closely with both Singer and Attenborough before moving to the USA, Peacock soon added his voice to the conversation to shape what all participants knew would be a landmark in television production, one which would concentrate a large share of the channel's

material resources over the next three years. Aubrey Singer's rough esti-
mate of the total cost for *Life on Earth* was above £700,000, and he was
strongly advised not to commission another major project before this one
was over.[21]

Peacock, who made his offer conditional on receiving the assurance that
Life on Earth would be Attenborough's 'only major television commitment
until production has been completed'[22] set high expectations for what
he reckoned would be a 'very expensive series'. The programme should
'delight the eye and engage the mind; … be authoritative and definitive
without being too academic and didactic for the general public; … be
visually fresh and exciting while avoiding the meretricious use of "pretty
film"; … be timeless without being dull. In short, it must be so good as to
seem effortless'.[23] Having joined the BBC the same year as Attenborough
(the two men were in training together; Chapter 3), Peacock shared with
Singer his concerns over the role Attenborough had carved for himself:

> I have a feeling that there is a danger of over using David speaking to camera,
> merely because he is so good at it! Seeing him wandering through a swamp
> or jungle is almost a personal trademark, and while very appropriate for TV
> explorations like "Eastwards with Attenborough", there was time in THE
> TRIBAL EYE when his piece to camera seemed to intrude upon the film.[24]

In other words, as Peacock put it to Rhodes, his fear was that Attenborough
would rely too much on a tried recipe, 'the Zoo Quest walkabout style',[25]
when the scope and ambition of *Life on Earth* called for a less conservative
approach. He also felt that Attenborough had not yet risen to the challenge
of the project, he was 'too didactic' and failed to provide viewers with a
'broader perspective—the "grand design" of nature, evolution or whatever'
against which they could see 'the beauty and diversity of life on earth' and
fill in 'at least some of the how's and why's'. To try and make his own

[21] John Stringer, 'Life on Earth', memo to Controller, BBC2, 17 October 1975, BBCWAC
T41/520/1.
[22] Controller-BBC2, 'Life on Earth', memo to Director of Programmes, Television, 24 October
1975, BBCWAC T41/520/1.
[23] Michael Peacock, personal letter to Aubrey Singer, 10 October 1975, p. 1. BBCWAC T41/520/1.
[24] Michael Peacock, personal letter to Aubrey Singer, 10 October 1975, p. 1. BBCWAC T41/520/1.
[25] Michael Peacock, personal letter to Mick Rhodes, 16 October 1975, BBCWAC T41/520/1.

perspective clearer, Peacock pointed Rhodes to Jacques Monod's *Chance and Necessity* (1970), 'the best approach to the how and why of evolution that I know of'.

Peacock's expectations about technical quality and presenting style met with approval at the NHU. The Bristol producers wanted 'everything to look fresher and better than before and we are particularly anxious to encourage the latest specialised film techniques to this end'.[26] Attenborough himself, it turned out, 'felt that talking to camera should be at a minimum'; a view Parsons shared, to whom 'any to camera words should be included only because they form part of a continuous narrative which simultaneously involves a piece of appropriate action for David'. But when it came to content and engagement with the sciences, the NHU producers were much more critical of Peacock's views. Taking up his 'grand design point', Rhodes, a trained ecologist, objected that 'the whole point with evolution is that there is no grand design. However, there are ways in which it *always* works (this is what Darwin was about) and so one will appreciate the grand perspective.'[27]

> I think what has happened is that David's treatment emphasises the taxonomic grouping of animals while John Sparks' earlier paper emphasised the functions. Somewhere in between is where we want to go.

As soon as the co-production arrangement was agreed at the end of November 1975, Aubrey Singer insisted that the special production unit set up for the series be established. Mick Rhodes had very strong ideas on who should work on the project and why. In a memo worth quoting extensively because it provides an unfiltered and vivid view on key characters in this story, Rhodes forcefully argued the case for his choice:

> John [Sparks] is, in my opinion, the person I most want as the third producer on "Life on Earth", … for I am trying to put together a balanced team.

---

[26] Mick Rhodes, 'Letters from Michael Peacock', memo to Controller BBC2, 22 October 1975, BBCWAC T41/520/1.

[27] Mick Rhodes, 'Letters from Michael Peacock', memo to Controller BBC2, 22 October 1975, p. 3. BBCWAC T41/520/1 (emphasis in original).

Chris [Parsons] is the expert organiser and editorialiser: he is also a very filmic film-maker. Richard [Brock] is the expert at getting the last dodo onto film, but he's not a filmic film-maker. He can go anywhere and overcome any odds to achieve rare and difficult footage—but he's not an editorialiser or organiser. David [Attenborough] is the charming performer and contact man with the wide world of zoological people, but he's not an abrasive mind (or indeed a stylish modern film-maker!).

What I need to add to this slightly bland mix are two things: irritant, grit and fast flowing possibly eccentric, ideas—plus a heavy dose of high quality academic zoological knowledge. John is a great provider of grit and ideas—but no great judge of them, (Chris/David will select/discard!). He also has the high academic knowledge. (He is, after all, the only television producer I know with a zoological Ph.D. behind him.). This team, it seems to me, balances well.

It could be that somewhere there is an alternative to John with even more attributes but I know of *no-one* with his combination of film-making experience, free-thinking attitude *and* zoological knowledge. No, he is quite definitely my first choice as the third leg of the stool.

And apart from anything else, he is at least 50% responsible for the idea in the first place![28]

By March 1976, the three producers (Parsons, Sparks and Brock), their assistants, a pool of researchers, two film crews, a film editor and his assistant, as well as a unit organiser, responsible for the day-to-day running of the structure and keeping an eye on the budget, were in place. By this stage, *Life on Earth* was recognised as 'a programme undertaking of unusual dimensions, which is going to make major demands—both physical and professional—on all those involved for a long period', as the internal advertisement for film crew positions read.

Everyone concerned must be possessed of a positive urge to get involved, to contribute and to be continuously and fully identified with the series.

---

[28]Editor, Natural History Unit, 'Life on Earth: John Sparks', memo—Staff Private—To Tom Williams, 13 November 1975, BBCWAC WE17/56/1 (original emphasis).

Naturally there will be a call for a high level of professional skill and expertise: this is taken for granted. But of almost equal importance will be the
question of personal attitudes and motivation within the team.[29]

Life on Earth originated in the professional culture of wildlife broadcasting which had accreted over the previous decade a blend of film-making
expertise and the ability to keep abreast of cutting-edge zoological knowledge. But *Life on Earth* should also be viewed as its final formative episode,
for it led to the creation, for Attenborough, of a new professional identity within the ecology of wildlife television production: wildlife television
writer and performer, or 'telenaturalist' (Gouyon 2011). Performing this
science-informed natural history in public, on television, and through the
production of television programmes, the telenaturalist embodies the idea
that wildlife television production is a way of practicing this brand of
natural history whilst promoting it. However, the addition of this professional identity to the existing professional culture of wildlife television
was not obvious and demanded some adjustments on the part of all those
involved.

Life on Earth as Television's Own Genuine Statement on Evolution

Soon after the start of production, in May 1976, dissensions began to
appear over what the series should be about and the division of labour
in the production unit. These resulted from some criticisms Singer and
Peacock had voiced after seeing Attenborough's first draft of some scripts.
To Singer, although they demonstrated 'scholarship', they had 'no authority', and were evidence that Attenborough had not yet appropriated the
concept of the series, that he was not identifying with it. Authority, Singer
suggested, could come from 'a sense of presence, and David Attenborough's relationship to the "out of vision" sequences is a crucial decision
that has to be taken'. Above all, Singer added, 'We have to give … a sense

---

[29] Gerry Bloomfield (Manager Programme Services, Bristol), '"Life on Earth"—Film crew and editing
staff', memo to All film cameramen & assistant film cameramen, all film recordists and assistant film
recordists, all film editors and assistant film editors, 17 February 1976, p. 2. BBCWAC SW3/32/1.

of fun which this slightly text-book approach seems to lack.'[30] In a further memo, Singer questioned the work conducted at Bristol more bluntly: 'I am still worried about the lack of original statement. What sort of really expert biological opinion are you canvassing apart from David?'[31] The series' co-producer, Michael Peacock, echoed Singer's concerns.

> I must confess to a general concern about David's approach. Even when I lean over backward to make allowances for the fact that the scripts are still in the first draft, I still feel that they are *lacking in perspective*. David makes little attempt to develop connections, whether scientific or philosophical, which would give new dimensions to his admirably lucid, descriptive narrative. His material seems to lack a sense of direction. We don't feel that we are part of an exciting scientific enquiry.[32]

These criticisms precipitated a crisis in the production unit, where it soon appeared that a 'considerable difference'[33] existed between the NHU's concept of the series and that of their script writer, David Attenborough. As Mick Rhodes explained at length to Singer,

> Most of the problem comes from conflict between David's view of the series and ours. I have always seen these programmes as being the intellectually stimulating—and definitive—story of evolution. David has a simpler view and tends toward some kind of very superior 'Zoo Quest' approach. There have already been several conflicts between David on the one hand and Chris and the producers on the other hand about precisely this point. These confrontations have produced some movement in David's position but not much. And David is being very dogmatic.[34]

More than just the next natural history television series, *Life on Earth* was a defining project, reaching deep down into the existential consciousness of

---

[30]Controller BBC2, 'Life on Earth', memo to Mick Rhodes, 6 July 1976, BBCWAC T41/520/1.

[31]Controller BBC2, 'Life on Earth', memo to Editor Natural History Unit, 23 August 1976, BBCWAC T41/520/1.

[32]Michael Peacock, personal letter to Mick Rhodes, 6 October 1976, p. 1. BBCWAC T41/520/1.

[33]Anonymous [Mick Rhodes], n.d. [end 1976?], 'Life on Earth', p. 1. BBCWAC T41/520/1.

[34]Editor, Natural History Unit [Mick Rhodes], 'Life on Earth', memo to C BBC2 [A. Singer], 9 September 1976. BBCWAC T41/520/1.

the NHU in Bristol and its producers, first and foremost Christopher Parsons. When at an early planning stage Aubrey Singer questioned whether the project, given its scale, ought to be conducted in Bristol at all, Tom Salmon, the regional senior manager based in Bristol and overseeing all production activities in the South-West, forcefully defended the NHU. He reminded Singer that 'historically, "Life on Earth" was conceived (and this was long ago) as being the big, classic, prestigious N.H.U. series. There has been endless talk about it, hopes of it, and pride in it.'[35] By the early 1970s, *Life on Earth* had become part of the NHU's professional identity and was intrinsic to how producers envisioned the future of the Unit. Subsequent discussions amongst the production team at the NHU mobilised the protagonists' deep beliefs about the form wildlife television should take, and its content.

Central to the discussions taking place at the beginning of the production phase was Attenborough's role as a scriptwriter. To him, scriptwriting did not involve research beyond the canonical literature with which he was already familiar. Rhodes, Parsons and their colleagues, took a different approach:

> During the summer of 1976 we realised that the research that we believed David would do in connection with his scriptwriting was not happening, and would not happen. This was partly because David did not wish to consult anyone before committing his own thoughts to paper and partly because his concept of the series did not demand it.[36]

This disagreement boiled down to the question of whether *Life on Earth* should be an original statement about evolution and the origin of life on the planet—in other words, whether broadcasting ought to get involved with research-in-the-making on these topics. Attenborough envisioned *Life on Earth* more as what Mick Rhodes described at the time as 'Glossy FE [further education]',[37] an exhaustive and visually pleasing synthesis of

---

[35]Acting Head of Network Production Centre, Bristol [Tom Salmon], 'Life on Earth', memo to C. BBC-2, 29 July 1974, BBCWAC T41/520/1.

[36]Anonymous [Mick Rhodes], n.d. [end 1976?], 'Life on Earth', p. 1. BBCWAC T41/520/1.

[37]Editor, Natural History Unit, 'Life on Earth', memo to C BBC2, 9 September 1976, BBCWAC T41/520/1.

the latest accepted text-book knowledge. By contrast, the NHU producers were in no doubt that *Life on Earth* should venture into uncertainty:

> One of the main problems is that Paleontological research is widespread and scientific papers come out each day which propose new theories, or challenge those recently published. Almost no-one agrees on anything. It is one thing to do a neat 50 minutes package on the Symbiotic Theory, but another to put that theory in perspective within our series. To do so, one needs to talk to far more people than one would normally be able to do in researching for a single 50 minuter. But here our 2 ½ year span does allow this and we should be in a good position to make *our own assessment* of overall scientific opinion immediately before the final narration is written.[38]

To NHU producers, wildlife television had a role to play in sifting through scientists' competing claims to knowledge and providing viewers with a basis on which to make their own decision about what is true. From this standpoint, wildlife television was to be an arbiter in scientific debates and controversies as they happen, an intervention in cultural debates taking place on scientific topics. This ambition was coherent with the internal culture of the BBC as it had been developing since the late 1960s. Alongside the emergence of a professional culture of television broadcasting in the period, was an acute sense that, when at its technical best, television could produce its own original statements about the world. An early and forceful advocate of this take on the medium's cultural role was none other than Aubrey Singer, writing in 1966, when still the head of Science and Features:

> There are times when television acts in its own right, … when it uses its power of communication not merely to convey other people's images but rather to create out of its potentialities its own genuine statements. This is the television at which we in television have got to aim. When we do we can claim equal responsibility with those who create the values of society.

---

[38] Anonymous [Mick Rhodes?], n.d. [end 1976?], 'Life on Earth', p. 1 (emphasis added). BBCWAC T41/520/1.

With architects, authors, scientists, designers, film-makers, with all those who created and communicate original work.[39]

In the case of wildlife television, Singer's view placed broadcasters on an equal footing to scientists in producing natural knowledge. This amounted to self-styling the television institution as a necessary intermediary between scientists and non-specialists. In essence, the broadcasting institution was offering scientists a means to recruit public support for their claims to knowledge. From this perspective, *Life on Earth* was a scaling-up of the approach already underpinning the programme *Life* (1965–1968), likewise intended to enable audiences to witness scientific debates as they happened and to reach their own conclusions about them (Chapter 6). In both cases, television producers and scientists share a stake in the necessity to convince television viewers of the cognitive legitimacy of the claims to knowledge presented in programmes (Barnes 1978; Callon 1986; Latour 1987). Besides, in this configuration, scientists become interested in maintaining a close and continuous relationship with the television institution, to keep open this means of reaching to the public for continuous support for their work. And broadcasters benefit from an easier access to scientists and their research, making the preparation work for large series easier. As the outcome of *Life on Earth* suggests, scientists agreed on the value of this arrangement. In total more than 500 of them contributed to the series, many afterwards asking for copies to use as teaching aides (Parsons 1982).

Enter the Telenaturalist

Although a divergence of concept for *Life on Earth* existed early on between the NHU producers and Attenborough, the main misunderstanding came from the fact that no space yet existed for the role which Singer, Peacock, and to some extent Rhodes and the Bristol producers encouraged Attenborough to carve for himself in the ecology of wildlife television production. Neither quite a producer nor a cameraman-director, Attenborough

---

[39]Aubrey Singer, 1966, 'Television: Window on culture or reflection in the glass?', *The American Scholar, 35*(2), 303–309.

had to fashion for himself a hitherto non-existent professional identity, that of wildlife television writer and performer, or telenaturalist, whose practice of natural history takes place in public, on television and through the production of television programmes. In the *Zoo Quest* series two David Attenboroughs had co-existed: (1) a character in the story, appearing in the films shot on location, and (2) a character addressing audiences, live from a studio in London. *Life on Earth* brought these two incarnations together in storylines which stage the presenter in the field.

Between the mid-1950s and the early 1970s, the division of labour at Bristol had evolved from presenter centred to film centric. At first, with *Look* (1955–1968), amateur naturalist cameramen brought films to the NHU, where a producer working with Peter Scott constructed a half-hour programme that involved Scott in a studio delivering an ad-lib commentary on the film. In this configuration, Scott's live performance was the central motif of the programme on which the producer could act, the film being mostly a support for it. With *The World About Us* (1967–1982), under Attenborough's impulsion, film became the major output for wildlife television, eclipsing the studio-based magazine and the associated figure of the presenter. In this configuration a producer would agree on a treatment for a story with a wildlife cameraman who then would go in the field and shoot. Using this raw footage, a film editor would then assemble the feature in collaboration with the producer who finally wrote the commentary. Although the producer could ask the cameraman-director to cover specific aspects, the onus of researching the story and finding interlocutors when necessary rested on the latter. Producers could occasionally get involved in some filming to complete the list of shots needed to tell the story visually, as for instance Parsons did when working with OSF on their *World About Us* film about the wildlife of Jamaica. But the producer's job mostly consisted in expertly evaluating the film material cameramen generated and its suitability to construct television programmes. As Jeffery Boswall was explaining to Gerald Thompson in 1967 (Chapter 8): 'I am employed as a producer. And if material offered is not of suitable content, in my judgement, then I must say so.'[40]

---

[40]Jeffery Boswall, personal letter to Gerald Thompson, 17 February 1967, p. 3. BBCWAC WE13/1,071/1.

The creation of a professional identity for the telenaturalist brought the presenter back, but with a twist, for it implied that producers were to lose their position at the top of the production chain. The executive producer for the series was Christopher Parsons. But as Mick Rhodes had perceived early on, Attenborough, as 'the storyteller', had to hold 'the casting vote' as the 'executive editor' for the whole series.[41] Scriptwriting was thus about creating the setting for his own on-screen, storytelling performance. As Mick Rhodes explained to Aubrey Singer in July 1976, Attenborough's scripts 'had come straight out of [his] head and library'.[42] Each was 'really meant to be a basic "information" treatment as a starting point for the production team to work up'. The task of researching the series' content fell back on the producers. Equipped with the episode outline, they could start 'a full scale research job … which involves a fair amount of travelling and talking'. Following this extensive period of research, the producers then fed back their findings to Attenborough to revise his earlier treatment into 'a genuine production script'. The refashioning of the ecology of wildlife television to accommodate Attenborough's new role as natural history writer and performer—as telenaturalist—placed him on top of the programme production chain.

Producers at the NHU had to adopt the working method developed notably in the BBC Science and Features Department, where a *Horizon* producer would spend six weeks researching a topic before filming an episode. The man heading the unit in Bristol, whom Attenborough had appointed precisely because of his *Horizon* background,[43] was to facilitate the producers' conversion to this redefinition of their professional culture. Indeed, in September 1976, Mick Rhodes informed Singer that he was moving 'much closer to the series' to assist with the script production.[44] By the start of 1977, Rhodes had got his unit in marching order and the filming campaign had begun in earnest. Several specialist cameramen were

[41]Anonymous [Mick Rhodes?], n.d., 'Life on Earth, a proposal for a major series from the NHU', p. 1. BBCWAC T41/520/1.

[42]Editor, Natural History Unit, 'Life on Earth', memo to C. BBC2, 21 July 1976, BBCWAC T41/520/1.

[43]Interview with author, 25 November 2015.

[44]Editor, Natural History Unit, 'Life on Earth', memo to C. BBC2, 9 September 1976, BBCWAC T41/520/1.

already busy filming around the world 'inundating [Bristol] with material, some of it extremely good'.[45] With most of the scripts for the series in 'mark 2 versions', filming for Attenborough's on-camera appearances could take place. A very busy year lay ahead, which would take the presenter from the Galapagos to East Africa through to a grand tour of America in the summer, before visiting Orkney and Durham 'followed by a mammoth trip from the Bosphorus or Gibraltar via Alaska to Australia from 22 August through to the first week of November', ending up with South America.[46]

The main feature of Attenborough's travelogue style, with which he first experimented in the *Zoo Quest* series, was his 'leap-frogging', appearing in one place and then the next as the narration progressed, a presenter's equivalent of the jump cut. As one location could be relevant to the narration in different episodes, all the on-camera appearances happening there had to be filmed at the same time, hence the necessity of having all the scripts ready at once. The editing of the opening of the first episode is quite deliberate in its use of such leap-frogging. Attenborough's first appearance is in a rainforest, clad in his then customary beige safari suit. First looking up a tree with binoculars and then crouching to observe a column of leaf-cutting ants, it is a direct quotation from the *Zoo Quest* series. The sequence is meant to reconnect viewers with the figure of the expert naturalist fashioned in the jungles of Sierra Leone, Guiana and Borneo (Chapters 3 and 4). Next, introducing Darwin's concept of natural selection in the Galapagos, pointing at the giant tortoises found there, and then meditatively sitting on a volcanic shore surrounded by sea lions, a moored ship in the background, Attenborough sets himself to retrace the great Victorian naturalist's intellectual journey, following in his footsteps. There he establishes Darwin as his forebear—and himself as a televisual Darwin. Finally, in his third appearance, Attenborough goes further, picking up things where Darwin had found them and introducing the methodological cornerstone of the series: observation. Scientists, Attenborough explains off-camera, use radioactivity to measure the age of

---

[45]Editor, Natural History Unit, 'Progress report: Life on Earth', memo to C. BBC2, 11 February 1977, BBCWAC T41/520/1.

[46]Editor, Natural History Unit, 'Progress report: Life on Earth', memo to C. BBC2, 11 February 1977, BBCWAC T41/520/1.

rocks and fossils. But a simpler way of dating rocks and fossils exists, 'that anyone can use, and there is no more dramatic place to do so than in the Grand Canyon in the American West'.[47] Coming early in the series, this commentary positions *Life on Earth* as exemplary of an enterprise of exploration of the natural world other than, but also equal to, science, based on one methodological precept: observation. Attenborough, going down the Grand Canyon as if travelling back in time, getting closer to the origin of life on Earth the deeper he goes, demonstrates his gift for observation, determining the age of fossils without the help of scientists' radioactivity. These on-camera appearances, carefully staged,[48] did not simply link together geographically distant parts of the story to make a coherent narrative whole but were key moments when the presenter, unconstrained by time and space, could freely roam the world and explain the order of things, expertly performing natural history on-screen and becoming part of the story.

The most powerful of these performances are those when Attenborough got 'dramatically involved with his subject'.[49] Undoubtedly the most memorable one, shot by Martin Saunders in Rwanda, in January 1978, shows Attenborough, in the penultimate episode of *Life on Earth*, physically interacting with wild gorillas in the jungle, a young one sitting on top of him. Originally only shot 'to show the lads back home', this unscripted encounter was eventually judged in Bristol so spectacular that it ended up edited in the relevant episode.[50] This sequence, more than any other, contributed to cementing the notion that Attenborough was at home in the wild and that his expertise sprang from an intimate knowledge of animals. This sequence evokes the earlier embraces in *Zoo Quest* between Attenborough and the female chimpanzee Jane, or the young orang-utan Charlie. Such close physical proximity with wild animals is a trope of the culture of natural history in the twentieth century, from Cherry Kearton

[47] Commentary *Life on Earth* Episode 1, 'The infinite variety' (BBC 1979).

[48] The descent down the Grand Canyon was filmed on the way up. For each shot the train of mules had to be turned backwards so that it looked like they were going down (Parsons 1982: 335).

[49] Christopher Parsons, 'Life on Earth', memo to Controller BBC-2, 2 February 1978, BBCWAC T41/520/1.

[50] J. Burgess and D. Unwin, 1984, 'Exploring the living planet with David Attenborough', *Journal of Geography in Higher Education*, 8(2), 93–113, 109.

to Attenborough and beyond, and is itself borrowed from the Victorian culture of big game hunting (Chapter 3). But it is also a hyperbolic version of the kind of encounters with wild animals showed in Eric Ashby's programmes (Chapter 5). The difference here is that the telenaturalist does not need to conceal himself, as the amateur naturalist cameraman did. Instead, wild animals come to him of their own accord. The telenaturalist can move freely from human society to the wilderness and back again, bringing to the metropole expert knowledge of nature.

Documenting the Making of *Life on Earth*

Just as Attenborough conceived of *Life on Earth* as a means of cementing his reputation at home and abroad—he famously refused any contract that would replace him with another presenter for the American version of *Life on Earth* (Burgess and Unwin 1984)—producers at the NHU understood the series as a defining moment in the Unit's history. It was to demonstrate wildlife television's ability, when at its technological best, to produce an original statement on the natural world. As the release date was approaching, the idea of documenting their achievement in a MOD emerged. By the end of 1977, as the final editing had begun, little doubt remained amongst all those involved, that 'the potential [was] here for an outstanding series'.[51] Rhodes, Parsons and Derek Anderson, the *Life on Earth* unit organiser responsible for administering the series' budget, started discussing the production of *The Making of Life on Earth*. First, Anderson approached Michael Peacock at Warner Bros, who agreed to contribute an additional £4000 to help finance this MOD. Envisaged as a thirty-minute trailer to advertise *Life on Earth*, this making-of documentary intended to give 'insight into the making of the series'[52] would emphasise the Unit's expertise.

To ensure that it was going to strike the right tone, and not come out as self-aggrandising, Rhodes and Parsons commissioned a director

[51]Controller BBC-2, 'Life on Earth', memo to Mick Rhodes, 31 January 1978, BBCWAC T41/520/1.

[52]Senior Producer, General Production Unit [Michael Croucher], 'Life on Earth special', memo to Editor NHU, 31 January 1978, BBCWAC WE17/53/1.

and an executive producer from outside the NHU, Colin Godman and Michael Croucher, respectively, both from the General Programme Unit at Bristol. Croucher had developed a specialty of producing documentaries about 'ordinary people' and their occupations. Rhodes's idea was to 'sublet [the NHU's] editorial rights completely' thereby placing the *Life on Earth* staff 'into the position of contributers [sic.]—like the management of the factory about which somebody is making a film'.[53] To Croucher and Godman, taking their cue from Rhodes's 1972 *The Making of a Natural History Film*, the MOD should not so much concentrate on the production of *Life on Earth* as use it as a pretext to shed light on the Bristol unit as a whole:

> The BBC may be criticised for looking at its own Natural History Unit, much as the BBC looked at Oxford Scientific Films some years ago. The self-examination in this case, however, is not a form of indulgence but rather recognising the opportunity to capitalise on the inevitable excitement to be generated by the showing of 'Life on Earth'.[54]

To celebrate, the NHU was to highlight wildlife television professionals' expertise at recreating nature on-screen. An early outline for the MOD shows that half of the film was to be dedicated to showing cameramen at work, and to hearing them 'talk of their work and problems. The time and patience required will be emphasised which will lead into the *necessity* of controlled filming to intercut with "authentic" footage.'[55] The projected MOD was thus envisaged as a means of justifying the hands-on approach to camera work in light of the greater efficiency at representing nature which filming under controlled conditions afforded film-makers. These disclosures about filming techniques were to work as selling points to promote the series:

---

[53]Editor NHU, 'The making of Life on Earth', memo to Derek Anderson, 24 February 1978, BBCWAC WE17/53/1.
[54]Michael Croucher and Colin Godman, 'The making of Life on Earth', memo to Mick Rhodes, 2 May 1978, pp. 7–8. BBCWAC WE17/53/1.
[55]Anonymous, 'The making of Life on Earth', BBCWAC WE17/53/1.

The aim of this 30 minute film is to provide the viewer with an insight of what was involved in the making of the most complex wildlife television series ever made. … This film will therefore attempt to convey this whilst involving viewers in rather more details on a number of biological filming problems—a formula that proved very popular in 'The Making of Natural History Film'. In doing this, we hope to provide an additional and tantalising promotion for the series.[56]

Imagined as an advertisement for *Life on Earth*, this MOD would have juxtaposed finished sequences from the series with footage depicting how they had been obtained. It was thus as much envisaged as being about providing evidence of the film-makers' skills as it was about serving as a means of educating audiences in a specific way of viewing the series, focusing their attention on the material practice of film-making as a source of knowledge of the natural world. By April 1978, Croucher and Godman had shot five sequences illustrating 'the collection and photography of the extraordinary creatures'.[57] They showed Peter Parks (from OSF) collecting plankton in the sea off the Great Barrier Reef and operating his optical bench to film the microscopic life forms, and cameraman Rodger Jackman waiting for days for the adult male Chilean frog (*Rhinoderma Darwinii*) to release froglets from its mouth, a behaviour which *Life on Earth* producers claimed had never been seen before.

The Making-of Life on Earth was never finished, and never shown as a programme in its own right. Croucher's and Godman's interests, as producers of the MOD, did not meet with those of others involved in the production of *Life on Earth*, notably Attenborough's. The two film-makers, specialists in the production of social documentaries on communities, insisted that nobody should stand out as a 'hero' in the story. Attenborough, they said, would be 'seen as just another member of the "family"'.[58] Croucher and Godman thus hoped to film BBC crews at work in the field, both doing some location filming and shooting Attenborough's links. They

---

[56]Anonymous, 'The making of Life on Earth', p. 1. BBCWAC WE17/53/1.

[57]Michael Croucher and Colin Godman, 'The making of Life on Earth', memo to Mick Rhodes, 2 May 1978, p. 4. BBCWAC WE17/53/1.

[58]Michael Croucher and Colin Godman, 'The making of Life on Earth', memo to Mick Rhodes, 2 May 1978, p. 7. BBCWAC WE17/53/1.

also planned to cover the work undertaken in the cutting room, emphasising the monumental quantity of film generated for the series. All this
purposely shot footage was to be interspersed with extracts from the series,
supplemented with a voice-over commentary to provide the background
stories to the shooting. Here Croucher and Godman thought especially
of such striking sequences as that of the gorillas to deconstruct them for
the viewers and reveal what had deliberately been left out of the frame,
concealed from viewers. Yet, in May 1978, Godman had not managed to
figure out what Attenborough's contribution to the MOD would be.[59]

In parallel to the conversation on the content of the MOD between
Croucher, Godman and Rhodes, its role as an advertisement tool for the
series came up for discussion during meetings about the promotion of *Life
on Earth*. In April 1978, one such meeting took place in which Parsons
conveyed Attenborough's objections to the MOD being used before the
transmission of the series. He was not, however, opposed to the film being
used after the series had been shown, in association with the promotion
of the book he was writing based on *Life on Earth* and to be released at
the same time as the series.[60] In their outline for the MOD, Croucher and
Godman had acknowledged the 'dangers of "blunting the impact" of the
key sequences in the series'.[61] Attenborough may have been reluctant to see
such striking sequences as his encounter with wild gorillas unveiled outside
the context of his performance in a way that could lessen their dramatic
impact. By June 1978, the MOD project, proving too contentious an
issue, was cancelled.[62]

All the existing film work did not go to waste, however, as the five
sequences depicting specialised wildlife cameramen at work were eventually shown. But this happened in a context where Attenborough, in his
role as the front-of-house figure, controlled the disclosure. He appeared as
a special guest in the BBC1 programme *Wildtracks*, hosted by Tony Soper,

---

[59]Colin Godman, 'David Attenborough—"The making of Life on Earth"', memo to Con. & Fin. Ex., 9 May 1978, BBCWAC WE13/10/1.

[60]Anonymous, n.d., 'Life on Earth. Promotion', p. 2. BBCWAC WE17/57/1.

[61]Michael Croucher and Colin Godman, 'The making of Life on Earth', memo to Mick Rhodes, 2 May 1978, p. 6. BBCWAC WE17/53/1.

[62]Anonymous, n.d., 'Life on Earth: Minutes of promotion meeting held on Monday, June 5th 1978', p. 2. BBCWAC WE17/57/1.

scheduled after the first eight episodes of *Life on Earth* had been broadcast on BBC2. Transmitted in late afternoon on a Friday, *Wildtracks* was mostly aimed at children.[63] On the set, Soper interviewed Attenborough about the vagaries of filming wildlife, a conversation interspersed with the five sequences Godman and Croucher had shot. Attenborough first answered on-camera, sharing his own experience of working on *Life on Earth*, before delivering an out-of-vision commentary to the filmed sequences. In this format, rather than appearing as 'just another member of the "family"',[64] Attenborough is the *pater familias*, lifting the curtain on how the series was made. Master of ceremonies, he controls the performance, which includes choosing what to reveal and conceal about it (Morus 2006), instructing viewers on how to make sense of the series. Through their eventual use in a programme in which Attenborough appears as 'special guest', these making-of sequences became part of the process by which, through *Life on Earth*, Attenborough completes the fashioning of his identity as an expert on the natural world, begun with such series as *Zoo Quest*.

'Any University Would Give an Arm and a Leg to Be Able to Use Those Films for Teaching'

Life on Earth is a milestone in British wildlife television and its relation to science. Three years in the making, with a final budget exceeding £1,000,000, and having mobilised BBC resources far beyond the Bristol unit, it marks the beginning of the era of high-production-value blockbuster series, international co-production deals, the so-called blue-chip documentaries (Bousé 2000; Cottle 2004). Its legacy extends to the latest BBC wildlife mega-series, including the 2017 *Blue Planet 2*, and the 2018 *Dynasties*, both built on the same template elaborated four decades earlier, and increasingly self-referential (Richards 2013). As Attenborough

---

[63] *WildTrack*, with Tony Soper, BBC1, 2 March 1979.
[64] Michael Croucher and Colin Godman, 'The making of Life on Earth', memo to Mick Rhodes, 2 May 1978, p. 7. BBCWAC WE17/53/1.

acknowledged, since *Life on Earth*, he has kept ploughing the same fur-row.[65] John Sparks echoed this view, commenting that all the series that followed *Life on Earth* 'have simply been expanding on each of the pro-grammes'.[66] But *Life on Earth* also established wildlife television's part-nership with scientific research to produce knowledge about the natural world. As a memo discussing ways to advertise the series indicates, this was a feature which Bristol was keen on widely publicising:

> It seems to me that we may have a sufficient number of sequences in which behaviour has been filmed for the first time to make a separate publicity sheet. This would be of particular interest to biological and general science magazines such as New Scientist. In addition, there are one or two events which are new to science—notably the birth of *Dasyuroides byrnie* (Kowari). I don't think it would be too difficult to make a significant list of others, e.g. *Cyclorana platycephalus* (Desert frog) discarding its cocoon underground; *Rhinoderma Darwinii* giving "birth" to froglets, etc.[67]

These were not empty claims by Parsons and his colleagues. As soon as the series was out, scientists approached whomever they knew at the BBC to try and get hold of copies of at least a couple of episodes of the series. A zoologist in the Zoology Department at the University of Edinburgh contacted Judy Copeland, married to Peter Copeland, the sound engineer responsible for the final mix of *Life on Earth*, to convey her department's thorough enjoyment of the series and her eagerness to obtain prints of some episodes to use as teaching material. It was *the* topic of conversation in the Department the day after each episode had been screened.

> I suppose Christopher Parsons realises that any university would give an arm and a leg to be able to use those films for teaching … What is the cost of buying individual programmes? I think the Department would probably

[65] *Life on Air*, talk at the BFI (London), 4 November 2015.

[66] John Sparks, Oral history interview, 11 July 2008, Wildscreen.

[67] Editor, Natural History Unit, 'Life on Earth Press launch', memo to J. Sparks, M. Salisbury, R. Brock, N. Cleminson, 19 October 1978, BBCWAC WE17/57/1.

buy two or three … for my course, the amphibian, marsupial and primate ones would be particularly wonderful to have.[68]

With *Life on Earth*, the NHU cemented its partnership with scientists, demonstrating the value of film-making to research. Camera work but also post-production, from editing to dubbing, could contribute to the production of new knowledge of the natural world. Even music had an epistemic role to play in so far as it enhanced visual perception. To viewers querying the use of music in the series, Parsons insisted that 'in *Life on Earth* we have used various production techniques, some of which we believed are enhanced by music'. The NHU producers had come to this conclusion after trying two pilot programmes on several sample audiences. These viewers' reactions had confirmed that 'for the majority of viewers, well-written commissioned music contributes much to the total enjoyment of a programme'.[69] A key role for music, as David Attenborough explained shortly after the release of *The Living Planet* (1984), was to evoke sensations beyond the realm of the visual to create as vivid a viewing experience as possible. 'Suppose there is a slightly threatening situation—in reality. You can't convey the temperature, or the wind, or the smell—all of which, in reality may contribute to that particular threatening atmosphere. But you can heighten that atmosphere on film with the judicious use of music.'[70] Another role for music was to bring coherence and a feeling of continuity to the programmes (Parsons 1971). In *Life on Earth*, this was notably helpful in relation to Attenborough's presentation style: his leap-frogging. The presenter, jumping from one continent to the next, put side-by-side geographically unrelated places and life forms. This resulted in programmes which, visually, were a succession of discrete vignettes, and whose only narrative thread was the presenter's continuous commentary, on- and off-camera. But even this seemingly disjointed structure, the outcome of intense editing sessions in the cutting room,

---

[68] Judy Copeland, 'Life on Earth', memo to Chris Parsons, 1 October 1979, BBCWAC WE17/57/1 (original emphasis).

[69] Christopher Parsons, reply to F. A. Lipscombe, *Radio Times*, 'Letters', 10–16 February 1979, p. 61.

[70] J. Burgess and D. Unwin, 1984, 'Exploring the living planet with David Attenborough', *Journal of Geography in Higher Education*, 8(2), 93–113, 103.

enhanced the series' epistemic power, for it showed broadcaster's ability to recreate nature on-screen.

In an interview to promote a repeat of the series on BBC1, Attenborough emphasised the epistemic value, for this 'glorious explanation of Darwin's theories of evolution', of his leap-frogging.

> We were able, for instance, to put together views of living amphibians which no one had been able to see in that range of time ever. No zoo could show you that amount. The visual effect was devastating. It had the same effect on me [Attenborough] as it did on everyone else. I remember the first time I saw the amphibian programme. I was speechless. My jaw was sagging with wonder.[71]

Collections of natural history specimens, for naturalists, are tools to produce knowledge because they allow for comparison. In the words of the eighteenth-century French naturalist George Cuvier, they enable naturalists 'to roam freely throughout the universe', providing them with an 'overview of the natural order as a whole' (Outram 1996: 261). Television spectators sitting in front of the screen are like naturalists standing in front of an open drawer in the calm enclosure of their study. *Life on Earth*, spreading before viewers a wide sample of related living organisms, enabled them, like Attenborough himself, to freely roam the world, transforming every domestic sitting room into an armchair naturalist's study.

Attenborough's comment in this interview, casting his method of script writing and storytelling as the televisual equivalent of natural history enquiry, is at the same time a display of modesty, which removes him from the picture and further asserts his status as a truth teller. Historian of science Steven Shapin noted that one feature of the development of experimental science in seventeenth-century England was to place a premium on experimenters' displays of modesty: 'A man whose narratives could be credited as mirrors of reality was a "modest man"; his reports should make that modesty visible' (Shapin 2010: 101). Attenborough, referring to the register of awe to describe his feelings upon seeing his own programme, removes himself from the production process as secondary to it. But in

---

[71] N. Wapshott, 1980, 'The perfect teacher, back with the animals', *The Times*, 1 March 1980, p. 14.

so doing, he delegates the epistemic power of his narrative technique to the series editing. In turn, this mechanical process comes to characterise the knowledge originating the series as an objective one, for it is defined as the product of the mechanical devices constituting the film-making apparatus, and not the subjectivity of individuals.

As a mechanically produced objective representation of the natural world, *Life on Earth* sits squarely in the cultural space of science. But it also promotes a definition of science as a public good, which belongs to all and in which, thanks to the television institution, the BBC, everyone can participate. Further into his interview with *The Times*, Attenborough continued:

> I can't tell you how touching some of the letters were. We were receiving about 100 a day. They came from children eight years old and professors of zoology. One professor wrote: 'But above all, I must thank you for reminding me why it was that I became a zoologist 50 years ago.'[72]

Such anecdotal evidence characterises wildlife television as the genuine heir of the original 'spirit of scientific enquiry'.[73] Cast as a new kind of collection of natural history specimens, produced through the mechanical means of film-making, *Life on Earth* sanctions the entry of the BBC's NHU in the cultural space of science as a legitimate participant in the production of knowledge of the natural world. The two following major series the NHU produced, *The Living Planet* (1984) and *The Trials of Life* (1990), each expanded on one of the two claims laid in *Life on Earth*. The first one, *The Living Planet*, further exemplified the epistemic power of film-making as a technical means of producing objective knowledge of nature. *The Trials of Life* stood as evidence of wildlife television's epistemic status as an equal partner in the scientific endeavour alongside scientists. Both of these series also had a making-of documentary as their last, thirteenth, episode.

---

[72]Wapshott, 1980.

[73]Desmond Hawkins, 1962, 'The BBC Natural History Unit. Report by the head of West Regional Programmes', p. 7. BBCWAC R13/462/1.

Self-Confident Experts

The Making of the Living Planet (1984) implemented the ideas discussed earlier in relation to *The Making of Life on Earth*, but it also firmly asserted the prominence of the series presenter, the telenaturalist, in the division of labour. Presented by humourist Miles Kington, the forty-minute documentary opens with the introduction of David Attenborough, dressed up for the field in his trademark khaki, sitting at a desk, presumably writing a script—the telenaturalist at work.

> One thing that distinguishes men from other living creatures is that only men make films about other living creatures, and perhaps one of the most famous and interesting of these film-makers is the species known as David Attenborough. … for this he has the necessary boundless curiosity and endless energy. What he doesn't have is the vast quantity of money and expertise that only the BBC can offer. He enjoys this rather strange, symbiotic relationship with the BBC, an odd and apparently friendly organism, whose workings we do not yet fully understand.[74]

To jokingly make David Attenborough the object of the MOD is to further downplay his agency in the production, thereby reinforcing his claim to objectivity. To frame the MOD as a pastiche of a wildlife film gives it a humorous edge which can be interpreted as yet another case of filmmakers presenting themselves as modest individuals, hence as truth tellers. One way for a natural philosopher to appear as a modest man was, for instance, to present himself as 'a drudge of greater industry than reason' (Shapin 2010: 101). *The Making of the Living Planet* introduced the use of discarded footage, bringing forth the drudgery of wildlife film-making. For example, the final section of the MOD, an interview with David Attenborough, is punctuated with footage from the successive takes of one sequence. Each time, Attenborough unsuccessfully tries to deliver a commentary on the kakapo, a nocturnal ground-dwelling parrot from New-Zealand, but has to stop because the bird is moving out of the frame. The repetition is comical, an effect reinforced by the spectacle of Attenborough growing increasingly frustrated with each attempt. The humorous

---

[74] *The Making of the Living Planet* (BBC 1984).

self-deprecating tone pervading *The Making of the Living Planet* can be analysed as a means for film-makers to secure credibility. Humour is a rhetorical strategy to provide viewers with '(inadequate) grounds freely to withhold their assent, and hence permitting them freely to constitute the basis of the assent they ultimately [have] to give' (Shapin 1994: 223). Humour enables film-makers to command their audiences' assent to the claim that the film-making apparatus is a legitimate tool to produce valid knowledge of the natural world. To show accidental footage of failures suggests to viewers that the artifice of film-making is not used to deceive. In other words, the playfulness of the MOD is intended to defuse potential criticism of, and in effect to normalise, the use of artifice in order to represent nature.

The Making of the Living Planet brings to the fore and legitimates the constructed nature of the wildlife television series, disclosing the tricks employed to represent nature on-screen. It is very complete in that respect, as it shows every aspect of programme making. Notably, a sequence explaining how views of the bottom of the Pacific Ocean were obtained highlights the role of music, composed by Elizabeth Parker, in creating not so much the illusion but the allusive atmosphere (Grimaud 2006) that will create the impression that the camera really captured views of the deep, when it was in fact filming a mock-up in a studio. The whole of the film-making apparatus is thus shown to contribute to creating the conditions whereby viewers can obtain general knowledge of the natural world. Rather than a cause for epistemic disqualification, artifice is presented as the necessary tool for wildlife film-makers to create on-screen the phenomenon they want to represent. In this respect, wildlife film-makers do not just show nature; they make it visible. Disclosure of the way this is achieved enables them to create an epistemic space for their practice alongside that occupied by science and to lay claims to expertise for themselves within that space by demonstrating their mastery of the technical means of film-making. *The Living Planet* (BBC 1984) was the achievement of self-confident wildlife broadcasters, who could afford to reveal some of their trade secrets without risk of undermining the trust they solicited from their audiences. On the contrary, the MOD for the series celebrated film-makers' expertise at reconstructing nature on-screen for the benefit of viewers. Standing as the embodiment of this new expertise, at the same time in and out of the story, the master of illusion bridging between this reconstructed version of nature and the real world is the telenaturalist.

The Abidance of the Telenaturalist

By the mid-eighties, wildlife broadcasters in Britain enjoyed the standing of professionals whose expert handling of the film-making apparatus enabled the creation of compelling television programmes which scientists could regard as worthwhile contributions to scientific knowledge—a long way from the amateur naturalist cameramen of the 1950s. Yet, as research for *The Trials of Life* (BBC 1990), the next major Bristol output, was underway, scientists began sending signals that their goodwill toward the NHU was not unlimited. If wildlife film-makers were to be considered as equal participants in the knowledge production endeavour, they had to acknowledge scientists' contribution in the production of television programmes. Receiving copies of programmes to use as teaching aides was no longer enough, as Peter Jones, who was the executive producer for *The Trials of Life*, soon found out:

> As we contacted some leading academics, there was an occasional adverse response. In getting to the bottom of this, I found that some felt that the financial aspects of the major series was all too much in favour of the BBC and David without any benefit (not even credits let alone financial) to the ultimate source of the ideas—the academic researchers.[75]

The push to align wildlife television with science, which Attenborough had initiated and nurtured in the mid-1960s as controller of BBC2, had made broadcasters depend increasingly on scientists for ideas and approaches that allowed them to keep filming animals in the wild without seeming to repeat themselves. In the mid-1980s, behavioural ecology, a renewed approach to the study of animal behaviour in the field, was developing out of Tinbergen's and Lorenz's ethology. Just like ethology, behavioural ecology rests on scientists' ability to recognize individuals among a population.

---

[75] Peter Jones, interview with author, September 2016. Unless mentioned otherwise, quotes thereafter are from this interview.

For behavioural ecologists try to understand how variations in individuals' behaviour, morphology, and physiology affect their adaptation to their environment and their reproductive success (Caro 1998: 14).

Peter Jones was the editor of the *The Natural World* series, which, in 1982, had replaced *The World About Us* (1967–1982) as the NHU's flagship programme. Jones was one of the *Horizon* producers who transferred to Bristol after Mick Rhodes had been placed at the head of the NHU. Several of his films for *Horizon* had dealt with natural history topics, and had enabled him to establish an extended network amongst scholars researching animal behaviour and evolution, which in turn led him to capture, early on, the birth of the new discipline in several episodes of *The Natural World*.[76] Cambridge zoologist Tim Clutton-Brock, with whom Jones had worked on the *Horizon* episode *The Red Deer of Rhum* (1978), had become one of Jones's regular science consultants after he'd started *The Natural World*. 'I had learned from Tim that many young researchers had entered this field of study and their work was creating almost a renaissance in behavioural studies in the field—or more accurately, the new paradigm, behavioural ecology.' Inspired by these conversations with Clutton-Brock, and others, Jones had used *The Natural World* as an outlet to popularise the new discipline, putting promising newcomers in charge of producing episodes on aspects of animal behaviour. Seeing these programmes combining high-quality photography with cutting-edge thinking from behavioural ecology led Attenborough to approach Jones in November 1986. In a short note, he expressed the wish to work on a twelve-part series with the young producers responsible for these episodes, each looking at one aspect of the life history of animals. Attenborough by then was an established presence in wildlife television. He had fronted two major successful series and was becoming the voice of nature on British television, being under contract with the NHU to record the commentary on each episode of the weekly programme *Wildlife on One*, 'the BBC One half hour series which was the proving ground for emerging young wildlife producers in Bristol'.[77] Attenborough offered Jones the post of executive

---

[76]Among Jones's notable natural history *Horizon* films are *The Selfish Gene* (BBC 1976); *The Human Animal* (BBC 1977); *The Red Deer of Rhum* (BBC 1978).

[77]Peter Jones, personal communication.

producer of what would be his next big series, *The Trials of Life*. This production exemplifies the tight collaboration between wildlife broadcasters and scientists.

Soon after he had agreed on the outline of the series with Attenborough (in January 1987), Jones offered it as a programme idea to the head of the NHU, then John Sparks, who took it to the controller of BBC1, Jonathan Powell, who immediately commissioned it (launching the search for co-production funding). Jones's first hire for the project was a researcher, Nick Upton. Fresh from his Ph.D. at Cambridge, 'Nick had just moved … to a Post-Doctoral Fellowship at Oxford, but on the basis of his brief experience of specialist macro-filming was prepared to give up his fellowship to come to Bristol for a short-term contract of a few years'. Soon the two men began researching the topic. The first stage was to send a questionnaire to university departments and active field researchers, as well as journal editors, asking them to report on arresting stories of animal behaviour, providing striking illustrations of the different stages of a life history: mating, feeding, home making, reproduction, parental care, and so on. The replies fed into a database, assembled using a cheap clone of an IBM PC and WordPerfect Library as a highly effective data entry system. The best stories extracted from this database then served as leads for the two men to go around the world visiting university departments and talking to researchers.

> Typically, later that year, I can remember visiting the University of Zurich (with Nick) and meeting Rudiger Wehner one morning to look into navigation by desert ants; and then, meeting Prof Hans Kummer for a brief lunch, I was astonished to hear that he had just reviewed the work of his brilliant young primate researcher Christophe Boesch who was observing highly cooperative chimp hunting in the Tai Forest. Alas, we just missed Christophe as he had set off back to the Tai forest, but on our return to Bristol, I debriefed the team on all our research efforts to find that our newest recruit Alastair Fothergill was very keen to follow up the Zurich stories.

The findings of this extensive phase of research were then presented to Attenborough, going through his preliminary list of programmes and distributing the stories gleaned from field researchers to each of them. In

keeping with the division of labour painfully hammered during the production of *Life on Earth*, during this meeting 'David … specifically said
that we should not attempt to generate any kind of treatment, and waving
his arms to indicate his extensive library behind him, he suggested we
should confine ourselves to researching and supplying ideas but leave him
to provide the scripts'. The writer-presenter was to remain free to insert his
own performances within the animal behaviour stories discovered during
the research. The producers for each episode and the unit manager worked
'with David in analysing and breaking the scripts down into sensible and
affordable filming trips combining his on-camera appearances with the
specialist wildlife elements', each script turning into a bespoke scenario
for the telenaturalist.

As part of the research process, Jones had taken his half-a-dozen-strong
team to the meeting of the Association for the Study of Animal Behaviour
(ASAB) in Madison, Wisconsin.

> This proved to be an extraordinary meeting with Plenary and other sessions,
> including Poster Sessions providing us with close to 50 or 60 ideas (about
> ten a day was the "strike rate" leading to a tremendous sense that the
> behavioural field work was indeed going through a remarkable period of
> new findings—much evident in Nick's research entries at that time).

During the Madison meeting, Jones sat with a senior group of academics
in order to talk over the feeling researchers had expressed about a lack of
recognition for their participation in such large BBC endeavours as *The
Trials of Life*. He explained that most of the money allocated to this type
of project was meant 'to cover the costs of extremely expensive location
filming'. He acknowledged 'that it was indeed their published/reported
work that was portrayed in the series, and without it we would lose an
invaluable and frankly irreplaceable contribution' but also insisted that
wildlife broadcasters needed to 'put all [their] money on the screen' in
order to meet audiences' expectations about the technical quality of the
programme. Otherwise, the BBC would cut the budget for this kind of
programme, not to mention the fact that 'the best series reflected very
well on contributing academic departments, both as regards reputation
and recruiting'. To compensate the researchers who gave their time in the

field to assist in the filming, Jones promised to explore ways in which they could be credited. This took the form of a special making-of programme to be broadcast as the thirteenth episode of the series 'because individual programmes presented solely by David could not feature the field researchers who were essential in capturing often new behaviours'.[78] Attaching an MOD to the big series became necessary to sustain the conception of wildlife television as a one-man show developed since *Life on Earth*, and revolving around the telenaturalist's solo performance in the field.

This MOD, *Once More into the Termite Mound* (BBC 1990),[79] established a model for the way in which wildlife broadcasters cast their relationship with scientists, further asserting the epistemic standing of wildlife film-making and using scientists' participation as a way of bringing epistemic legitimacy to film-makers' necessary use of artifice. The stories told in *The Trials of Life* originate in the turn to individualization, which took place in studies of animal behaviour in the 1970s–1980s with the development of behavioural ecology.[80] Behavioural ecologists advising television producers in the field could point them toward clearly identified individual animals, which could then be built as characters in the stories. Yet in the late 1980s, individualisation of animals was still seen as one of the unfortunate proclivities of wildlife film and television and a slippery slope toward anthropomorphism. To acknowledge scientists' contribution in the series highlighted at the same time that when it comes to observing animal behaviour in the field, scientists and film-makers shared a common approach to wild animals. On the basis of this shared methodology, it became easier to claim that scientists and film-makers participated in the same enterprise.

Structured around a series of interviews conducted by David Attenborough, *Once More into the Termite Mound* is punctuated with footage of scenes from the shooting of the series. All interviews, save the last

---

[78] Peter Jones, interview with author, 29 June 2015. Jones added in further exchanges: 'We also financially compensated researchers when they took time off their own work to be fully committed to assisting *The Trials of Life* team - an agreeable daily rate that was fair if not all that generous!' (personal communication to author).

[79] Producer: Michael Gunton; Executive Producer: Peter Jones.

[80] Interview with Peter Jones, 1 December 2014.

one, are with field scientists, whose work was essential to key scenes. These interviews, intended to disclose the role of scientific advisors on the set, foreground the value of individualising animals for understanding their behaviour but also filming them. In turn, the detailed knowledge of animals obtained through individualisation is identified as what will ultimately preclude anthropomorphising them, the main pitfall of individualisation. For instance, to a question on the 'danger' of anthropomorphising elephants, once she has started individualising and naming them, pachyderm expert Cynthia Moss explains that the more we know about elephants, thanks to studying individuals, the more difficult it becomes to think of them as human or as possessing human attributes. Progressing from being a source of error, individualisation becomes epistemically appropriate and heuristically fruitful. In these interviews, scientists demonstrate that they share the cornerstone of their methodology with film-makers who, by implication, appear as co-participants in the scientific enterprise through film-making.

The final interview in the MOD, based on an idea from Attenborough, hammers this point home. It features wildlife cameraman Paul Atkins and takes place in a BBC cutting room. Atkins was the cameraman responsible for shooting the iconic sequence showing killer whales snatching seal pups on a Patagonian beach. In the interview, he describes how he obtained underwater close-ups of a killer whale, whilst describing his relationship with the killer whale expert who advised him on-site. The story ends with both the cameraman and the scientist getting in the water and filming an orca up-close. The cameraman tells of the scientist being transfixed by wonder at being able to see 'his' animals in their element:

And he was just exhilarated at having finally seen his whale underwater. That was Mel, a male that he's watched for seventeen years, just watching the back and the dorsal fin and bursting out of the water to feed, but he had never been that close to the animal.[81]

The whole story belongs in the register of the adventurous explorer's tale. Told from the cameraman's perspective, it presents him as prone to taking

---

[81] *Once More into the Termite Mound* (BBC 1990).

risks and putting his and the scientist's life in danger to obtain shots that will enrich audiences' visual experience of wildlife. Putting one's life on the line stands, here, as evidence of trustworthiness. Early wildlife film-makers, like Cherry Kearton, had similar stories to tell about their close encounters with dangerous animals when risking their lives to bring close-up shots of tigers or lions to the metropole (Gouyon 2011). In this story, participating in film-making becomes a practice that enriches scientists' experiences, bringing them closer to their object of study to get new knowledge of animals they have watched for several years. From this vantage point, wildlife film-making is more than just the communication of scientists' results; it is a genuine participation in the scientific enterprise as it contributes to scientists' research work. The same theme was played out in *Making Waves*, the MOD for *The Blue Planet* (BBC 2001). Here, the collaboration between a scientist and wildlife cameramen for the filming of a sequence on killer whales is described as having allowed the former to observe in full a behaviour she knew existed but had never seen for herself in her seventeen years of observation. Here again, the scientist's participation in film-making is presented as a means of expanding her knowledge, as much as that of everybody else (Gouyon 2016).

The production of *The Trials of Life* and the way it was discussed in the related MOD is located within the paradigm around which the professional culture of wildlife television operates. In this paradigm, a large production team works in close collaboration with scientific advisors to create a set of filmed narratives revolving around one character: the telenaturalist. The only visible human presence, he imparts his worldview; audiences see through his eyes. Even when absent from the screen, he tells the story; spectators are to listen.[82] This paradigm came into being with *Life on Earth* and then became consolidated with the next two series, *The Living Planet* and *The Trials of Life*. And just as all the subsequent major blockbuster series coming out of the NHU since then have been variations on the themes first explored in these three series, they are also based on this paradigm structured around the figure of the telenaturalist, performed by David Attenborough.

---

[82]The telenaturalist is similar to Andrew Tudor's 'telexpert' (Tudor 1981).

One reason for this continuity is to be found in the generation leap between the first two series and *The Trials of Life*. Both *Life on Earth* (1979) and *The Living Planet* (1984) had been produced by individuals who had already accumulated often two decades of experience in wildlife broadcasting before working on these projects, which represented a climax in their career. By contrast, the production team of *The Trials of Life* included a few debuting producers, who then applied the modus operandi developed in the 1990 series to subsequent programme making. One of them was, notably, Alastair Fothergill, initially hired as an assistant producer and promoted to producer during the production. Fothergill went on to become head of the NHU from 1992 to 1998 and produced several other big Attenborough series, from *Life in the Freezer* (BBC 1993) to *Our Planet* (Netflix 2019) and through to *Blue Planet* (BBC 2001) and *Planet Earth* (BBC 2006), either as a BBC producer, or more recently, as the co-director and co-founder of Silverback Films, a production company specialising in wildlife. Fothergill is exemplary of a generation of wildlife film-makers who began their career when the paradigm structured around the telenaturalist was solidified and contributed to perpetuating it because they perceived it as a tried and tested formula for successful programme making.

The endurance of the telenaturalist from *Life on Earth* to *Our Planet* also rests on the fact that the most prominent incarnation of the role remains Attenborough, who created it. Wildlife television broadcasting is a collective endeavour. And as this book has tried to highlight, many, from Desmond Hawkins to Christopher Parsons, have contributed to shaping it with their ideas about nature and aspirations for the medium, as a powerful tool to know and appreciate the world we inhabit. Yet, it seems that one individual, David Attenborough, through a combination of chance and design, shaped the medium and how it operates in a way that enabled him to stand, for the last four decades, at the top of the pyramid of wildlife documentaries production, a position exemplified by his solo screen performance as the telenaturalist. For, although the telenaturalist is a fictional character who only exists in the narrative universe defined by the film, the role is at the same time played by David Attenborough as himself, existing in the real world and making it quite difficult for anybody else to be a telenaturalist as well. The question remains of whether the telenaturalist will become wildlife television's equivalent of Dr. Who, different performers taking up the role, or whether the role will disappear when the age of Attenborough ends, others inventing new roles for themselves in a renewed approach to wildlife television.

References

Barnes, B. (1978). *Interests and the growth of knowledge.* London: Routledge.

Boon, T. (2017). 'Programmes of real cultural significance': BBC2, the sciences and the arts in the mid-1960s. *Journal of British Cinema and Television, 14*(3), 324–343.

Bousé, D. (2000). *Wildlife films.* Philadelphia: University of Pennsylvania Press.

Burgess, J., & Unwin, D. (1984). Exploring the living planet with David Attenborough. *Journal of Geography in Higher Education, 8*(2), 93–113.

Callon, M. (1986). Éléments pour une sociologie de la traduction: la domestication des coquilles Saint-Jacques et des marins-pêcheurs dans la baie de Saint-Brieuc. *L'Année Sociologique, 36,* 169–208.

Caro, T. M. (1998). *Behavioural ecology and conservation biology.* Oxford and New York: Oxford University Press.

Cottle, S. (2004). Producing nature(s): On the changing production ecology of natural history TV. *Media, Culture and Society, 26*(1), 81–101.

Gouyon, J.-B. (2011). From Kearton to Attenborough: Fashioning the telenaturalist's identity. *History of Science, 49*(1), 25–60.

Gouyon, J.-B. (2016). 'You can't make a film about mice just by going out into a meadow and looking at mice': Staging as knowledge production in Natural History Film-making. In M. Willis (Ed.), *Staging science* (pp. 83–103). London: Palgrave Pivot.

Grimaud, E. (2006). Têtes multiples et jeux d'optique: Ou l'art de truquer les dieux hindous. *Terrain, 46,* 85–106.

Latour, B. (1987). *Science in action: How to follow scientists and engineers through society.* Cambridge: Harvard University Press.

Monod, J. (1970). *Le Hasard et la Nécessité.* Paris: Editions du Seuil.

Morus, I. R. (2006). Seeing and believing science. *Isis, 97*(1), 101–110.

Outram, D. (1996). New spaces in natural history. In N. Jardine et al. (Eds.), *Cultures of natural history* (pp. 249–265). Cambridge: Cambridge University Press.

Parsons, C. (1971). *Making wildlife movies, an introduction.* Newton Abbot: David & Charles Ltd.

Parsons, C. (1982). *True to nature.* Cambridge: Patrick Stephens.

Richards, M. (2013). Global nature, global brand: BBC Earth and David Attenborough's landmark wildlife series. *Media International Australia, 146*(1), 143–154.

Shapin, S. (1994). *A social history of truth.* Chicago and London: The University of Chicago Press.

Shapin, S. (2010). *Never pure: Historical studies of science as if it was produced by people with bodies, situated in time, space, culture, and society, and struggling for credibility and authority.* Baltimore: Johns Hopkins University Press.

Tudor, A. (1981). The Panels. In T. Bennet et al. (Eds.), *Popular television and film* (pp. 150–158). London: BFI, The Open University Press.

10

Afterword

Until only quite recently, and for most of the age of Attenborough, environmental issues have been the elephant in the room when it comes to BBC Natural History programmes. The Bristol unit's engagement with the loss of biodiversity, environmental destruction, or the climate emergency, often judged too timid in the past decades, led to growing controversies and producers parting ways with the BBC. Most notably, Richard Brock, who produced *The Living Planet* (Chapter 9), left the NHU out of frustration about the treatment of environmental issues in wildlife programmes: 'So the frustration for me was this stuff wasn't getting out there, nothing was being done about this. It was a con almost and it wasn't an organised con, it was just the way the system was working with controllers and, you know, their attitude to those sort of programmes.'[1] Or, as environmental activist George Monbiot, evoking the time when he worked at the NHU in the 1990s, revealed:

---

[1] Richard Brock, 2007, Oral history interview. Wildscreen.

© The Author(s) 2019
J.-B. Gouyon, *BBC Wildlife Documentaries in the Age of Attenborough*,
Palgrave Studies in Science and Popular Culture,
https://doi.org/10.1007/978-3-030-19982-1_10

In 1995 I spent several months with a producer, developing a novel and imaginative proposal for an environmental series. The producer returned from his meeting with the channel controller in a state of shock. "He just looked at the title and asked 'Is this environment?' I said yes. He said, 'I've spent two years trying to get environment off this fucking channel. Why the fuck are you bringing me environment?'"[2]

In this book, I have considered the establishment, over five decades, of wildlife film-making and television as epistemically productive ways of engaging with nature. The previous chapters tell the story of the emergence and consolidation of a specific paradigm for wildlife documentaries, culminating in the big, so-called Attenborough series, or blue chip documentaries. By way of conclusion, I want to briefly consider wildlife television's engagement with environmental topics. For, in the last years, it has led to a reappraisal of wildlife television's political role from neutral observer of wildlife to active participant in environmental debates. I will examine the timing of this shift in relation to changes in the life sciences' relationship to the environment, and the emergence of what has been called an 'endangerment sensibility' (Vidal and Dias 2016).

As I am writing this conclusion, David Attenborough, at 92, once more fronts a major series of the kind spawned by *Life on Earth*. But *Our Planet* is not a BBC series. It has been produced by Netflix, the web-based media content provider. To have the man who more than anyone else has come to embody the BBC's natural history output appear on what is seen, in 2019, as the main competitor of such traditional broadcasters as the BBC is a revolution in the wildlife television microcosm. What differentiates this Netflix production from its BBC equivalents is that it no longer ignores or minimises the threats facing all the habitats and animals on display. Since *Life on Earth*, Attenborough regularly brought up environmental concerns in his series (Richards 2013) but mostly did so in a final segment without ever dedicating a full major series to a discussion of the loss of biodiversity or climate change (with the exceptions of *State of the Planet* [BBC, 2000] and *The Truth About Climate Change* [BBC,

---

[2]George Monbiot, 'David Attenborough has betrayed the living world he loves', *The Guardian*, 7 November 2018. Available online at https://www.theguardian.com/commentisfree/2018/nov/07/david-attenborough-world-environment-bbc-films. Last accessed 3 April 2019.

2006]). In contrast to these BBC series, '*Our Planet* places clearer emphasis on the fragility and interconnectedness of all the species and eco-systems on display, and on the huge impact humanity has had on them in so short a time.'[3] This suggests that if BBC wildlife television is to remain relevant in the twenty-first century, it no longer can avoid the issue of the threat facing the environment and the animals it puts on display. Broadcasters will have to use their power to raise awareness of the environmental dire straits we are in through the production of politically engaged wildlife documentaries, or themselves face extinction.

Despite an association between film and the conservation of wildlife going back to the origins of wildlife film-making, BBC wildlife television's relationship with such pressing environmental issues as climate change, the loss of biodiversity, or environmental destruction is not straightforward. The very early advocates of the protection of endangered African wildlife used the film medium to proselytise, encouraging big game hunters to drop their guns and pick up a camera instead (Mitman 1999; Gouyon 2011). In the 1950s and early 1960s, conservationist Peter Scott used his series *Look* to sway public opinion toward supporting his conservationist agenda, with such episodes as *L for Lion* in 1961, or the more literally titled *Conservation in Action* (1964). Scott used *Look* to announce the launch of the WWF and then enrol television audiences in support of its work. After Peter Scott's efforts, though, Bristol waited until the first decade of the twenty-first century to fully engage in the environmental movement.

Along its fifty years of existence, wildlife television has co-evolved with the life sciences. Under the impulsion of David Attenborough, the Bristol-based wildlife broadcasters embraced ethology and fashioned themselves as participants in the production of knowledge about the behaviour of animals in the wild. This approach was in keeping with the then prevailing cultural understanding of nature as an entity existing principally to be explored, known and managed, which had emerged from the professionalisation of the life sciences from the end of the nineteenth century onward (Anker 2001). The framing of wildlife television according to this epistemologically driven understanding of nature found its expression in

---

[3]Lucy Mangan, 'Our planet review—Attenborough's first act as an eco-warrior', *The Guardian*, 5 April 2019. Available online at https://www.theguardian.com/tv-and-radio/2019/apr/05/our-planet-review-david-attenborough-netflix-eco-warrior-activist-bbc. Last accessed 5 April 2019.

the progressive marginalising of amateur naturalists at the same time as links between broadcasters and scientists were growing ever tighter. As the previous chapter showed, it all culminated in the production of milestones such as *Life on Earth* (1979) or *The Trials of Life* (1990), which their makers envisioned as conveying authoritative statements about the scientific understanding of life on the planet. Until the mid-1990s, concerns over the loss of biodiversity remained peripheral to the cultural space of the life sciences (Takacs 1996; Sepkoski 2016). Likewise, wildlife conservation, and more broadly any kind of environmental concerns, were only secondary to wildlife television's primary epistemic concern.

In the late 1990s and early 2000s, slightly different programme formats were tried in Bristol, taking inspiration from the soap opera genre, the main example here being *Big Cat Diary* (1996) and its sequel *Big Cat Week* (1998–2006). These series, structured around emotionally charged narratives, shaped animal protagonists into full-blown characters engaged in dramatic action. From the soap opera genre, they borrowed story hooks and cliff-hangers, putting emotions and storytelling to work to deliver a strong conservationist message (Richards 2014). These wildlife soap operas avoided straying too far from the notion that producing the programmes was at the same time to generate expert interpretations of empirical evidence of animal behaviour as captured by the cameras. But while staging presenters engaged in emotionally intense relationships with wild animals, these series at the same time promoted their political, emotional, and aesthetical feelings as relevant to the programmes' epistemic goal.

These programmes are concomitant with the rise in the life sciences of the biodiversity movement, which likewise proposed to make sense of natural diversity not just as an entity to be known through the expert interpretation of empirical evidence but also on an emotional, aesthetic, political, ethical and spiritual basis as a fragile and threatened one to be protected (Takacs 1996; Sepkoski 2016). This new valuing of biodiversity in the life sciences has been analysed as a consequence of the emergence of the 'endangerment sensibility': 'a complex of knowledge, values, affects and interests characterized by a particularly acute perception that some organisms and things are "under threat," and by a purposeful responsiveness to such a predicament' (Vidal and Dias 2016: 2). The rise of this mindset translated into calls to renew how research is done, to embrace moral principles and emotions as integral parts of knowledge production. In other words, from the mid-1990s onward, changes in discursive and scientific practices in the life sciences conflated conservation activism with the biological study of biodiversity. With the emergence of 'the endangerment sensibility', to engage in the scientific study of biodiversity has become a militant act (Vidal and Dias 2016). As the example of *Big Cat Week* demonstrates, producers working at the NHU have not been immune to this development. But these series about big cats in an African reserve, the Masai Mara, remained an oddity in the wildlife television landscape dominated by the BBC, and the closest Bristol would get to taking the endangerment turn. Wildlife film-makers wanting to take a more militant approach began operating outside the confines of the BBC.

In 2000, Richard Brock left the BBC's NHU after thirty-five years of a career begun with Desmond Morris's series *Life* (Chapter 6), during which he produced such landmarks as *The Living Planet* series. Concerned with the BBC's reluctance to address environmental issues,[4] he launched the Brock Initiative to produce locally oriented wildlife films in countries in the Global South and destined for local audiences. For example, in 2005, in collaboration with the charity Earthwatch, the Brock Initiative worked with local schools and communities in the East African Rift Valley to document, with fifteen short films, the environmental changes and ecological degradation in the region surrounding Lake Naivasha in Kenya, following its rapid economic development and population increase. Shown to local audiences, the films served as educational resources in the region to raise awareness about the local environment (Blewitt 2010: 101). The Brock Initiative is based on the assumption that films can only have a positive impact on local populations' relationship with the environment if the stories they tell use their language and environmental and cultural references. Only then will the films enable audiences to understand what is at stake and inspire them to act accordingly. In keeping with the endangerment sensibility, the key idea at the core of the Brock Initiative is that only by giving control of the camera to the populations most concerned with

---

[4] See for instance G. Monbiot, 2018, 'David Attenborough has betrayed the living world he loves', *The Guardian*, 7 November 2018. Available Online at https://www.theguardian.com/commentisfree/2018/nov/07/david-attenborough-world-environment-bbc-films. Last accessed 14 April 2018.

environmental issues will films have a positive impact in terms of wildlife and habitat conservation. One cannot help but remark that the approach championed by Richard Brock implicitly places the burden of environmental conservation on countries and populations which have most to suffer from the destruction of the biosphere but do not bear the main responsibility for the overexploitation of natural resources in the Global South, which mostly benefits the Global North. Yet, the Brock Initiative highlights the notion that to privilege the direct lived experience is essential to build a powerful documentary discourse about wildlife and environment conservation, for it all comes down to audiences choosing the world they want to live in as humans.

The endangerment sensibility, notes historian of environmental sciences Etienne Benson, 'seems inextricably tied to a vision of species as frozen in time and of conservation as an effort to insulate them against change' (Benson 2016: 191). In other words, such a vision of species and their conservation is a cinematic one, for film was first developed as a technology to capture movement and the world in its own image (Canales 2011). This idea that film could be put to work to freeze species in time and insulate them against change found an expression in the project imagined first by Christopher Parsons in the mid-1990s of a collection of film and photographs of all endangered species, which he called 'an electronic archive of living organisms' (Parsons 1996: 161). This project was eventually realised in the website Arkive.org, launched in 2003 and maintained online until 2019, under the aegis of the Wildscreen trust.[5] Arkive.org was intended to work as a tool for conservation, preserving, on film, species which are vanishing in the wild. A notion central to this web-based collection was that its films of animals could well become the only trace that remain of them if they eventually go extinct. Therefore, preserving the films was part of the effort to preserve wildlife. By developing an endangerment sensibility, it seems that scientists have adopted a cinematic understanding of wildlife. A question which now needs to be answered is whether the current understanding of scientific conservation and the valuing of biodiversity that goes with it is an outcome of the current generation of conservation scientists growing up with wildlife television, which over the

[5]https://www.wildscreen.org/.

past fifty years has striven to display the wondrous diversity of life on Earth.

References

Anker, P. (2001). *Imperial ecology: Environmental order in the British Empire, 1895–1945.* Cambridge, MA: Harvard University Press.

Benson, E. (2016). Endangered birds and epistemic concerns. In F. Vidal & N. Dias (Eds.), *Endangerment, biodiversity and culture* (pp. 175–194). London: Routledge.

Blewitt, J. (2010). *Media, ecology and conservation: Using the media to protect the world's wildlife and ecosystems.* London: Green books.

Canales, J. (2011). Desired machines: Cinema and the world in its own image. *Science in Context, 24*(3), 329–359.

Gouyon, J.-B. (2011). From Kearton to Attenborough: Fashioning the telenaturalist's identity. *History of Science, 49*(1), 25–60.

Mitman, G. (1999). *Reel nature: America's romance with wildlife on film.* Cambridge, MA: Harvard University Press.

Parsons, C. (1996). The electronic zoo and other components of Wildscreen world—A new approach to wildlife visitors attractions about the natural world. *Conference Proceedings, Trends in Leisure and Entertainment,* Maastricht, The Netherlands, 159–167.

Richards, M. (2013). Global nature, global brand: BBC Earth and David Attenborough's landmark wildlife series. *Media International Australia, 146*(1), 143–154.

Richards, M. (2014). The wildlife docusoap: A new ethical practice for wildlife documentary? *Television & New Media, 15*(4), 321–335.

Sepkoski, D. (2016). Extinction, diversity, and endangerment. In F. Vidal & N. Dias (Eds.), *Endangerment, biodiversity and culture* (pp. 62–86). London: Routledge.

Takacs, D. (1996). *The idea of biodiversity: Philosophies of paradise.* Baltimore: Johns Hopkins University Press.

Vidal, F., & Dias, N. (Eds.). (2016). *Endangerment, biodiversity and culture.* London: Routledge.

Index

© The Editor(s) (if applicable) and The Author(s) 2019
J.-B. Gouyon, *BBC Wildlife Documentaries in the Age of Attenborough*,
Palgrave Studies in Science and Popular Culture,
https://doi.org/10.1007/978-3-030-19982-1

Printed by Printforce, the Netherlands